SOCIAL LIFE
AMONG THE INSECTS

BEING A SERIES OF LECTURES DELIVERED AT THE
LOWELL INSTITUTE IN BOSTON IN MARCH 1922.

BY

WILLIAM MORTON WHEELER Ph.D. Sc.D.

PROFESSOR OF ECONOMIC ENTOMOLOGY, HARVARD UNIVERSITY; DEAN OF THE
BUSSEY INSTITUTION; HONORARY FELLOW AND RESEARCH ASSOCIATE,
AMERICAN MUSEUM OF NATURAL HISTORY.

NEW YORK

HARCOURT, BRACE & WORLD, INC.

JOHNSON REPRINT CORPORATION   JOHNSON REPRINT COMPANY LTD.
111 Fifth Avenue, New York, N. Y. 10003   Berkeley Square House, London, W. 1

To

My Son

RALPH EMERSON WHEELER

Ἴθι δή, ἦν δ' ἐγώ, τῷ λόγῳ ἐξ ἀρχῆς ποιῶμεν πόλιν. ποιήσει δὲ αὐτήν, ὡς ἔοικεν, ἡ ἡμετέρα χρεία. Πῶς δ'οὔ; Ἀλλὰ μὴν πρώτη γε καὶ μεγίστη τῶν χρειῶν ἡ τῆς τροφῆς παρασκευὴ τοῦ εἶναί τε καὶ ζῆν ἕνεκα. Παντάπασί γε.

"Then, I said, let us begin and create a state; and yet the true creator is necessity; who is the mother of our invention. True, he replied. Now the first and greatest of necessities is food, which is the condition of life and existence. Certainly."

— PLATO, *Republic* II, 11.

PREFACE

THE six lectures which form the present volume were given at the Lowell Institute in Boston, February 27th to March 16th, 1922. After appearing in several consecutive numbers of the " Scientific Monthly " (June, 1922 to February, 1923) they are here reproduced with only very slight verbal changes, although, owing to the rigid limitations of time, some of the paragraphs, especially in the second and third lectures, were omitted during delivery. I wish to express my grateful acknowledgment to Dr. J. McKeen Cattell for generously supplying the publishers with the blocks of the figures for use in the present volume, after their publication in his magazine.

The lectures were designed to present a sketch of our present knowledge of the behavior of the various groups of social insects. While preparing them I realized that the materials at my disposal were far too intricate and voluminous to permit of a well-balanced or symmetrical treatment within the limits of six lectures. I endeavored, therefore, to throw the emphasis on the fundamental nutritive *motifs* in the phylogeny, ontogeny, and maintenance of insect societies. This seemed to me to be the more necessary, because subjects like polymorphism, nest-architecture, sensory reactions, homing behavior, etc. have been adequately emphasized in the past, whereas ecologists, geneticists and physiologists, at least in this country, seem to be neglecting the very important rôle of nutrition, both in the behavior and structural modifications of animals.

The fourth and fifth lectures are a much condensed résumé of materials contained in my volume " Ants, their

Structure, Development and Behavior," Columbia University Press, 1910, with the addition of certain facts and considerations in which I have been interested during more recent years.

As an aid to the reader who may care to extend his studies of the social insects, I have added a documentary appendix, containing considerable bibliography and several notes on special topics which could not be adequately treated in the lectures. Since I have published a rather voluminous list of the ant-literature up to 1908 in the volume above mentioned, the bibliography of the third and fourth lectures is mainly restricted to a selection from the more recent literature.

The illustrations are derived from many sources. I am greatly indebted to a number of investigators for the generous loan of photographs and drawings, especially to Prof. I. W. Bailey, Dr. F. X. Williams, Prof. Carl Hartman, Mr. H. O. Lang, Dr. J. Bequaert, Mr. Etienne Roubaud, Mr. A. B. Casteel, Mr. F. W. L. Sladen, Mr. John Tee-Van, Dr. E. F. Phillips, Dr. O. S. Strong, Dr. Carlos Bruch, Mr. Charles Janet, Mr. Horace Donisthorpe, Dr. A. D. Imms, Prof. F. Silvestri, Mr. Thos. E. Snyder, Mr. Nathan Banks and Dr. Alfred Emerson. I am also indebted to the authorities of the American Museum of Natural History for the photographs reproduced in Figs. 53 and 54. As on many former occasions, Prof. C. T. Brues has greatly assisted in making clear reproductions of several figures from the monographs of recent authors and in correcting proof.

BUSSEY INSTITUTION

FOREST HILLS, BOSTON, MASS.
    DEC. 10, 1922.

CONTENTS

SOCIAL LIFE AMONG THE INSECTS

LECTURE I

GENERAL REMARKS ON INSECT SOCIETIES.
THE SOCIAL BEETLES

DURING the past fifty years, the science of living organisms has itself, like a living organism, developed so rapidly that it has more than once changed its aspect and induced its votaries to change their points of view. The future historian of the science will probably emphasize the difference of attitude towards the living world exhibited by Darwin and his contemporaries and that of the present generation of twentieth century biologists. He will notice that the works of the Victorians abound in such phrases as the "struggle for existence," "survival of the fittest," "Nature, red in tooth and claw," and disquisitions on the unrelenting competition in the development, growth and behavior of all animals and plants. This struggle, as you know, was supposed to constitute the very basis for the survival of favored forms through natural selection. There can be no doubt that even to-day we must admit that there is much truth in all this writing, but we would insist that it depicts not more than half of the whole truth. To us it is clear that an equally pervasive and fundamental innate peculiarity of organisms is their tendency to cooperation, or "mutual aid," as it was called by Prince Kropotkin.[1] Even to the great Victorian naturalists the fact was familiar — though they failed to dwell on its great social significance — that all living things are genetically related as members

3

of one great family, one vast, living symplasm, which, though fragmented into individuals in space, is nevertheless absolutely continuous in time, that in the great majority of organic forms each generation arises from the cooperation of two individuals, that most animals and plants live in associations, herds, colonies or societies of the same species and that even the so-called " solitary " species are obligatory, more or less cooperative members of groups or associations of individuals of different species, the biocœnoses. Living beings not only struggle and compete with one another for food, mates and safety, but they also work together to insure to one another these same indispensable conditions for development and survival. The phenomena of mutualism and cooperation are, indeed, so prevalent among plants and animals and affect their structure and behavior so profoundly that there has arisen within very recent years a new school of biologists, who might be called " symbiotists," because they devote themselves to the investigation of a whole world of microorganisms which live in the most intimate symbosis within the very cells of many if not most of the higher animals and plants.[2]

If asked why it seems advisable to devote six lectures to social life among the insects, I might say that these creatures exhibit many of the most extraordinary manifestations of that general organic cooperativeness which I have just mentioned, and that these manifestations have not only an academic but also a practical interest at the present time. For if there is a world-wide impulse that more than any other is animating and shaping all our individual lives since the World War, it is that towards ever greater solidarity, of general disarmament, of a drawing together not only of men to men but of nations to nations

throughout the world, of a recasting and refinement of all
our economic, political, social, educational and religious
activities for the purposes of greater mutual helpfulness.
As Edward Carpenter[3] says:
" The sense of organic unity, of the common welfare,
the instinct of Humanity, or general helpfulness, are things
which run in all directions through the very fibre of our
individual and social life — just as they do through that
of the gregarious animals. In a thousand ways: through
heredity and the fact that common ancestral blood flows
in our veins — though we be only strangers that pass
in the street; through psychology, and the similarity of
structure and concatenation in our minds; through social
linkage, and the necessity of each and all to the other's
economic welfare; through personal affection and the ties
of the heart; and through the mystic and religious sense
which, diving deep below personalities, perceives the vast
flood of universal being — in these and many other ways
does this Common Life compel us to recognize itself as a
fact — perhaps the most fundamental fact of existence."

The social insects may also be singled out for special
treatment for the following more particular reasons: first,
because they represent Nature's most startling efforts in
communal organization and have therefore been held up
to us since the days of Solomon as eminently worth imi-
tating, or to be avoided as an " abschreckendes Beispiel ";
second, because these organizations are simpler and more
perspicuous than our own and we can study their origin, development and decay and subject them to experimentation;
third, because many of them represent clean-cut products
of comparatively simple evolutionary tendencies and hence
final and relatively stable accomplishments; fourth, because they show us the extent to which social organization

can be developed and integrated on a purely physiological and instinctive basis, and by contrast therefore throw into sharper relief some of the defects and virtues of our own more intellectual type of society; and fifth, because they are so remote from us that we should be able to study them in an unbiased and truly scientific spirit.

I wish to dwell somewhat on the third of these reasons for the purpose of placing in clearer perspective the great antiquity and completeness of the social organization of insects. Some years ago the museums of Königsberg and Berlin sent me for study an extraordinary collection of ants in lumps of Baltic amber.[4] There were 9,560 specimens, representing 92 species and 43 genera. As you know, the Baltic amber is merely the fossil resin of pines which flourished during Lower Oligocene Tertiary times in the region which is now Sweden. The liquid resin exuded from the tree-trunks precisely as it does to-day, and great numbers of small insects, especially ants, were trapped in the transparent, viscid masses which hardened, fell from the trees or remained after the rotting of the wood and were carried down by the streams and embedded in what is to-day the floor of the Baltic Sea and the soil of Eastern Prussia. The lumps are now brought to the surface either by mining or by the action of the waves which cast them up on the beaches. So beautiful and lifelike are the insects preserved in the amber that by comparison all other fossils have a singularly dull and inert appearance. Many of the specimens which I was able to examine were as exquisitely preserved as living ants embedded in Canada balsam by some expert microscopist. My study showed conclusively that the ants have undergone no important structural modifications since the Lower Oligocene, that they had at that time developed all their various castes just as we see them to-day, that their larvæ and pupæ were the same, that they attended plant-lice, kept guest-beetles in their nests and had parasitic mites attached to their legs in the very same peculiar positions as in our living species, and that at least six of the seven existing subfamilies and many of the existing genera were fully established. Some of the species in the amber were even found to be practically indistinguishable from those now living in Northern Europe and North America. The Baltic amber also contains social bees, wasps and termites, and though these are not so well known, what I have said of the ants will also *mutatis mutandis* prove to be true of them. Since my work was published Cockerell and Donisthorpe have described a number of ants from the Bagshot Beds of the Isle of Wight, also of Oligocene age, and very recently Cockerell has described a typical ant, *Eoformica eocenica,* from even earlier strata, the Green River Eocene of Wyoming.[5] We must conclude, therefore, that these insects — and the same is very probably true also of the wasps, bees and termites — had their origin in the Cretaceous, if not earlier. What I wish to emphasize is the fact that all the main structural and social peculiarities of these insects were completed by the beginning of the Tertiary and that they have since been merely marking time or developing only the slight modifications which serve to distinguish genera, species, subspecies and varieties.

Now how many years have elapsed since the beginning of the Tertiary? Geologists have, of course, made many and diverse estimates. I shall take the most recent, which are much in excess of earlier computations. Barrell gives the time since the beginning of the Tertiary as 55 to 65 million years.[6] But the social insects are the most recent — the mere newly rich, so to speak — in the great class

Insecta, which has a fossil record extending back to the Upper Carboniferous. And as our earliest known fossils are perfectly typical insects, it is probable that the earliest Hexapods made their appearance in the Silurian, if not earlier. This would make the period during which these wonderful creatures have been living and multiplying on our planet about 300 million years!

In order that we and the impatient reformers in our midst may experience the proper feeling of humility let us now compare the age of man and his society with that of the ants. During the Oligocene and early Miocene, while these insects, together with the uncouth primitive mammals, represented the dominant animal life of the plains and forests of the globe, the early Primates were just splitting into two tribes, one of which was destined to produce the modern apes, the other the Hominidæ, or humans. Our ancestors were probably just forsaking that life among the tree-tops which, as F. Wood Jones has shown, has left its ineffaceable impress on all the details of our anatomy.[7] A large part of the diet of these early Hominids and their immediate ancestors probably consisted of those same ants which had already developed a cooperative communism so complete that in comparison the most radical of our bolsheviks are ultra-conservative capitalists. By a hundred thousand years ago our ancestors had reached the stage of the Neanderthal man, whose society was probably somewhat more primitive than that of the Australian savage of to-day. And so far as the actual, fundamental, biological structure of our society is concerned and notwithstanding its stupendous growth in size and all the tinkering to which it has been subjected, we are still in much the same infantile stage. But if the ants are not despondent because they have failed to produce a new social invention or convention in 65 million years, why should we be discouraged because some of our institutions and castes have not been able to evolve a new idea in the past fifty centuries? I find that social habits have arisen no less than 24 different times in as many different groups of solitary insects.[8] Careful investigation of the life-histories of tropical species will probably increase this number. These 24 societies, which I propose to consider in more or less detail in this and the following lectures, represent very different stages in the evolution of the social habit. Some of them are small and depauperate, mere rudiments of societies, some are extremely populous and present great differentiation and specialization of their members, whereas others show intermediate conditions. And while each of the 24 different societies has its own peculiar features, we nevertheless observe that all of them have arisen in the same manner and have the same fundamental structure. Each is a family consisting of two parent insects and their offspring or at least of the fecundated mother and her offspring, and the members of the two generations live together in more or less intimate, cooperative affiliation. During the long history of the Insecta this situation has developed time and time again and quite naturally out of the very general propensity of female insects to lay their eggs on food suitable to the hatching larvæ or to make protective structures or burrows, to store them with food and to oviposit on it. As a rule, the mother insect then dies and never sees her offspring, but all such parental care, which is also very prevalent among many other animals and even among plants, is nevertheless a potential or implicit nursing or fostering, which readily becomes actual or explicit in such species as manage to survive the hatching of their young and can therefore continue to feed and protect them. It is difficult, nevertheless, to draw a hard and fast line between certain solitary forms and some of the societies or families I have selected, for there is a finely graded series of cases of parental care between complete indifference to the offspring and the families of what may be called the incipiently social or subsocial forms. As the societies grow in size and complexity they naturally change from associations in which the progeny depend on their parents to associations in which the parents come to depend on their progeny.

John Fiske and others have claimed that human society has been rendered possible by a lengthening of infancy and childhood, since this obviously involves more elaborate care of the young by parents and a greatly increased opportunity of learning on the part of the child.[9] This is true, but it is equally true that the adult life of the parents must also be prolonged to cover the retarded juvenile development, and the insects show us that the lengthening of the adult stage comes first and makes social life possible. In solitary insects, of course, it is just the brevity of adult life that prevents the development of the social habit, no matter how long the larval period may be. This period may, in fact, extend over months or even years in certain insects which have an adult stage of only a few days or hours.

Momentous consequences necessarily follow from the lengthening of the adult life of the parent insect and the development of the family, for the relations between parents and offspring tend to become so increasingly intimate and interdependent that we are confronted with a new organic unit, or biological entity — a super-organism, in fact, in which through physiological division of labor the component individuals specialize in diverse ways and become necessary to one another's welfare or very existence. Since this integration necessarily leads to an important modification of the activities of the original solitary insects composing the society, it will be advisable to dwell for a few moments on the basic behavior of insects.

The activities of insects, like those of other animals, are an expression of three fundamental appetites or appetencies. Two of these — hunger and sex — are positive and possessive, the other — fear or avoidance — is negative and avertive. These appetites appear as the needs for food, progeny and protection. So far as I am able to see, they manifest themselves in insects in essentially the same manner as in the higher animals, such as birds, mammals and man. The appetites of hunger and sex arise from internal stimuli which compel the organism to make random or trial and error movements till appropriate, specific external stimuli are encountered. Then a sudden, consummatory reaction occurs and the relieved organism lapses into quiescence till the internal stimuli again make themselves felt. In the case of fear or aversion, harmful or disagreeable stimuli, usually of external origin, cause random movements till the organism escapes or succeeds in ridding itself of the noxious or discomfort-producing situation, when it becomes quiescent. And the behavior of insects, like that of other animals, seems to be made up of successions of such appetitive or avertive cycles, which may be repeated during the life-cycle, or — and this is particularly true of insects — the whole life-cycle may consist of a few appetitive cycles of very elaborate patterns — the so-called " instincts." [10]

Now when insects or other animals, for that matter, take to living in societies these fundamental appetites, which as

solitary individuals they have been exercising for millions
of years, are by no means lost or suppressed but become
peculiarly modified. Since the environment of the social
is from the outset much more complex than that of the
solitary insect, it must respond not only to all the stimuli
to which it reacted in its presocial stage but also to a great
number of additional stimuli emanating from the other
members of the society in which it is living. Even man, as
Berman says, " with the growth of his imagination and the
increase in number and density of his surrounding herd,
has become the subject of continuous stimulation." ¹¹ The
result seems to be a greatly increased responsiveness of the
organism. It becomes, so to speak, socially sensitized, and
all its appetites and emotions become hypertrophied or
even perverted. This will be clear for the insects from the
following very summary considerations:

1. Social life encounters serious and urgent difficulties in
the matter of food, for the colony must have access to a
supply which is abundant, nutritious and easily and continuously available in order that all the adult members
as well as the young may be adequately nourished. Such
an ideal food-supply is rare so that social insects are, as
a rule, chronically hungry and in the presence of food
positively greedy. Whenever possible both bees and ants
gorge themselves to the utmost. While an ant is feeding
on nectar or syrup her abdomen may be snipped off with a
pair of scissors, without interrupting her repast.¹² We
shall see, however, that she appropriates only a very small
portion of the swallowed food and that she distributes
most of it among her nest-mates. Hence, though she behaves like a glutton, we must refrain from regarding her
as such. When we see a man importuning everybody for

food or money we naturally regard him as avaricious or greedy, but when we learn that he is turning in all his collections to the Red Cross, he is transformed in our estimation. Not only does the social insect thus develop an unusual appetite for food but it also develops elaborate methods of apportioning the food among the adults and brood of the colony according to their various needs. Furthermore, the greatest economy in the use of food, which is of course energy, must be practised and various methods of preserving and storing it for consumption during seasons of scarcity must be devised. And since insect societies must compete with many other hungry animals they tend to specialize in their diet and to take to foods that can not be readily utilized by other organisms. All this specialization leads eventually to the development of a caste peculiarly adapted to provisioning the colony. As we shall see, this caste comprises the so-called "workers."

2. The reproductive gives rise to even more serious difficulties than the nutritive appetite. If all the individuals in the colony are permitted to reproduce without restraint, the population will soon outrun the food-supply and all its members will suffer from malnutrition or starvation, or it will have to resolve itself into smaller and feebler communities, and spread over a larger territory. The higher social insects have overcome this difficulty by rigidly restricting reproduction, except when food happens to be unusually abundant, to a few individuals and suppressing it in all others. Hence the fecundity of certain females, the queens, and of the males, or drones, becomes greatly enhanced or hypertrophied, while the remaining females, the workers, are reduced to physiological sterility. But it was found most convenient, while thus developing the queens and males as a reproductive and the workers as a

nutritive caste, and depriving the latter under normal conditions of the capacity for reproduction, to leave them in possession of their primitive parental instincts, that is, an ardent propensity for nursing the brood.

3. In the higher social insects fear is very readily aroused and can be easily studied in all its manifestations from abject cowardice and " death-feigning " in small and feeble species to panic rage in very populous communities. It is certainly of great biological significance, because these insects and their helpless brood are sedentary, or fixed in a particular environment, and are therefore exposed to the unforeseen attacks of enemies, inundations or great changes of temperature. Hence, we find that they not only make elaborate nests and fortifications but have developed powerful jaws, hard skulls, pungent or nauseating secretions and deadly stings. The workers originally assumed the protective rôle in addition to their other functions, but in many ants and most termites a special warrior or soldier caste has been evolved. Then, precisely as in man, many wasps, bees and ants found that the best method of defence is offence and their enemies were attacked before they could reach the nest. From this it was, of course, only a step to the organization of marauding and plundering expeditions and the development of aggressive warfare.

All the very complicated manifestations of the hunger, sex and fear appetites are so inextricably interwoven and interdependent that it is impossible adequately to study any one of them in isolation. I shall therefore have to refer to all of them again and again, but I wish to put the main emphasis in these lectures on the hunger appetite, because it is the most fundamental, exhibits the most astonishing developments, and is found to have an even greater influence on the reproductive and protective appetites than we had supposed. The recent work of the biochemists and physiologists on the vitamines and internal, or endocrine secretions, or "encretions" as some German investigators call them, has shown that extremely minute quantities of certain substances may have very profound and far-reaching effects on the metabolism, structure and functioning of living animals, and there has long been a suspicion that the differentiation of the fertile and sterile castes among social insects may be due to very delicate chemical stimuli. I shall endeavor to show that such stimuli may also play a determining rôle in maintaining the integrity or solidarity of many insect societies.

Before describing the various societies in greater detail I wish briefly to compare them with human society. I use this word in the singular, because at the present time, owing to the greatly increased facilities of transportation and intercommunication, what were once numerous independent human societies have practically fused or are about to fuse to form one immense, world-wide society. Human and insect societies are so similar that it is difficult to detect really fundamental biological differences between them. This assertion may be supported by the following considerations:

1. It is sometimes said or implied that human society is a rational association, due to intelligent cooperation, or contract among its members, whereas insect societies are merely physiological or instinctive associations. The second part of this statement is correct, but he who would seek support for the first part in the works of present day sociologists, psychologists and philosophers will be disappointed. The whole trend of modern thought is towards a greater recognition of the very important and determining

rôle of the irrational and the instinctive, not only in our social but also in our individual lives. The best proof of this statement is to be found in the family which by common consent constitutes the primitive basis of our society, just as it does among the insects, and the bonds which unite the human family are and will always be physiological and instinctive.

2. It may be said that insect societies are discrete entities, each of which arises as a single family, increases in population for some time and then dies away, whereas human society — the Great Society of Graham Wallas [13] — is a mixture of families and groups which grow and continue indefinitely. This is an important distinction but not absolute, since human society must have arisen from a single family or a few families, such as we find among the anthropoids. The difference would therefore seem to lie in the fact that our society no longer repeats its earliest phyllogenetic stage as does that of the social insects. But there are some insects, such as the honey-bee and some South American bees and wasps, that no longer repeat this incipient stage but from time to time send off new colonies, or societies by swarming, much as did the Phoenicians and early Greeks and the nations of western Europe in more recent times.

3. Korzybski, in an interesting book entitled " The Manhood of Humanity," has recently endeavored to emphasize another difference, the existence of social heredity, or what he calls " time-binding," in human society and its absence among animals. Certainly no one can overestimate the importance of tradition and social heredity. We should still be in the anthropoid stage if we had failed to preserve and add to the capital of culture and mores transmitted to us by former generations or ceased to trans-

mit them and the fruits of our own activities to our descendants. It is clear also that the social insects do not bequeath libraries, institutions and bank accounts to their posterity, and that each colony or society begins anew with the structural and instinctive equipment acquired by true, or organic heredity. This explains why we see so little change in these insects during the past 50 million years. Nevertheless, the distinction is not absolute. There are, as I shall show, ants, termites and beetles that cultivate fungi and bequeath them to succeeding generations. Social insects may also be said to bequeath real estate, that is, their nests, pastures and hunting grounds; and since the young queens of ants and termites often live for some time in the parental nests before they establish colonies of their own, there is reason to believe that they may acquire a very slight amount of experience by consorting with their sisters and parent queen.

4. It may be said that the social insects differ from man in not having learned the use of tools, but there are species of ants that use their larvæ as shuttles in weaving the silken walls of their nests, and the marvelous engineering feats of many social insects show that they are our close rivals in controlling the inorganic environment.

5. That they have acquired an equally astonishing control of their organic environment is shown by the fact that they are the only animals besides ourselves that have succeeded in domesticating other animals and enslaving their kind. In fact, the ants and termites may be said to have domesticated a greater number of animals than we have, and the same statement may prove to be true of their food-plants, when they have been more carefully studied.

6. It may be maintained that we have developed language, and this, of course, is a true distinction, if we mean

by language articulate speech, but the members of an insect society undoubtedly communicate wiah one another by means of peculiar movements of the body and antennæ, by shrill sounds (stridulation) and odors. The wonder has always been, not that there are so many differences in structure between such disparate organisms as insects and man, but that there are so many striking similarities in behavior. And the wonder grows when we find that social organization at least incipiently analogous to our own has arisen *de novo* on at least 24 different occasions in nearly as many natural families or subfamilies belonging to five very different orders of insects. A list of the groups that form these various societies is given in the accompanying table.

Coleoptera (Gynandrarchic)	1. Scarabæidæ (Copris, Minotaurus)
	2. Passalidæ (Passalus)
	3. Tenebrionidæ (Phrenapates)
	4. Silvanidæ (Tachigalia Beetles)
	5. Ipidæ (Ambrosia Beetles)
	6. Platypodidæ (Ambrosia Beetles)
Hymenoptera (Gynarchic)	*Sphecoidea*
	7. Sphecidæ (Sphex)
	8. Bembicidæ (Digger Wasps)
	Vespoidea
	9. Eumeninæ (Synagris)
	10. Zethinæ (Zethus)
	*11. Stenogastrinæ (Stenogaster)
	*12. Epiponinæ (Chartergus, Belonogaster, etc.)
	*13. Rhopalidiinæ (Rhopalidia, etc.)
	*14. Polistinæ (Polistes)
	*15. Vespinæ (Vespa)
	Apidæ
	16. Halictinæ (Halictus)
	17. Ceratininæ (Allodape)
	*18. Bombinæ (Bumble-bees)
	*19. Meliponinæ (Stingless Bees)
	*20. Apinæ (Honeybees)
	*21. Formicidæ (Ants)

Dermaptera (Gynarchic)	{	22. Forficulidæ (Earwigs)
Embidaria (Gynarchic)	{	23. Embiidæ (Embia)
Isoptera (Gynandrarchic)	{	*24. Termitidæ (Termites, or "White Ants")

In this list the first to tenth, the sixteenth and seventeenth, and the twenty-second and twenty-third are incipiently social or subsocial; the remaining ten, marked with asterisks, are definitely social. In the termites and all the beetle groups the colony consists of a male and female parent and their offspring of both sexes; in all the Hymenoptera, Dermaptera and Embidaria the female alone founds the colony, which is developed by her daughters. The former groups are therefore gynandrarchic, the latter gynarchic. These differences will become clearer as we proceed. Let us examine first the six beetle societies which have been developed by species belonging to as many different natural families.

1. *Scarabæidæ* — For our knowledge of the habits of the dung-beetles we are indebted to one of the greatest entomologists, J. H. Fabre.[14] His observations are recorded in parts of four of the ten volumes of his "Souvenirs Entomologiques," and comprise some of their most remarkable chapters. Some notion of the difficulties which he encountered while working out the life-histories of these insects may be gleaned from his statement that he did not succeed in completely elucidating the habits of one of them, the sacred Scarabæus, till he had had it under observation for nearly forty years. He studied quite a number of species and found startling diversity in their behavior. Some of them, the Aphodii, e.g., merely lay their eggs in fresh dung and the hatching larvæ feed on the substance. Among the

others, which resort to much more elaborate methods of
caring for their progeny, three different types of behavior
may be distinguished:

A. The Sacred Scarabæus (Fig. 1) above mentioned and
many allied forms are fond of the open sunlight and are
often seen making perfect spheres of fresh cattle manure
and trundling them away to cavities in the soil or under
stones. These pellets are devoured by the beetles. Fabre

FIG. 1.— Sacred scarabæi (*Scarabœus sacer*) trundling their pellet
of dung. (After E. J. Detmold.)

found that a single beetle will not only eat but digest a mass
of dung equal to its own body-weight in 12 hours. When
the female beetle is ready to lay she makes a very similar
pellet, but this time of sheep's dung, and rolls it into an
elliptical chamber which she has previously excavated in
the soil. This chamber is about as large as one's fist. She
then makes a crater-shaped depression surrounded by a circular flange at one pole of the pellet, lays a large egg in it

and draws the material of the flange over it till it is completely enclosed. The pellet is now pear-shaped. Thereupon the mother beetle leaves the chamber and proceeds
to dig another and provision it in the same manner. The
hatching larva consumes the inside of the pellet, pupates
within it and emerges as a beetle in due season. There is,
of course, nothing social about this insect. But in a
smaller, allied form, *Sisyphus schœfferi* (Fig. 2), Fabre

Fig. 2. — Male Sisyphus beetle (*Sisyphus schœfferi*) holding the
dung pellet while the female digs the burrow to receive it. (After
E. J. Detmold.)

found that the male helps the female trundle her pellet to
a convenient spot, guards it while she excavates a cavity,
assists her in lowering the pellet, waits for her till she has
oviposited in it in the same manner as the Scarabæus, and
then departs with her to repeat the performance.

    B. In Copris, of which Fabre studied two species, *C. his-
panus* and *C. lunaris*, we have a closer approach to a social
condition. These insects are crepuscular and dig a chamber

as large as a large apple immediately under the pile of dung. This is then carried down in masses and leisurely devoured. During the breeding season, however, the beetles associate in pairs and the male and female not only cooperate in excavating a chamber but also in nearly filling it with dung, which they then proceed to knead into the form of a smooth ellipsoid as large as a turkey's egg (Fig. 3). In the case of *C. hispanus* the male then deserts the female and the latter proceeds to cut the ellipsoid up into four

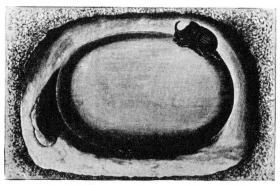

Fig. 3.—Spanish Copris (*Copris hispanus*) fashioning her large ellipsoid of dung in her subterranean chamber. (After J. H. Fabre.)

spherical pellets, each of which is treated like the pellet of the Sacred Scrarabæus, provided with an egg and converted into a regular ovoid (Fig. 4). The mother remains in the chamber guarding the pellets and keeping them free from fungus growth for four months, while the larvæ are developing within them. After the young beetles emerge the mother accompanies them to the surface of the soil and the family disperses. In the case of *C. lunaris* the male remains in the chamber with the female and helps her manufacture the ovoids, which owing to his assistance are twice as numerous as they are in *hispanus*. When the young beetles emerge they are escorted to the surface by both parents.

C. Other beetles, like Geotrypes, Onthophagus and Minotaurus, dig long tubular tunnels into the soil immediately under the dung. As a rule, they do not make spherical pellets but pack the deeper, blind end of the burrow with layers of dung till it forms a sausage-shaped mass above the egg or enclosing it at one end. The behavior of *Minotaurus typhœus* (Fig. 5) is even more astonishing than

Fig. 4. — Spanish Copris (*Copris hispanus*) guarding her ovoids of dung in the subterranean chamber. (After E. J. Detmold.)

that of Copris. The male and female beetles mate in March and together dig a tubular gallery straight down into the soil to the remarkable depth of five feet. The male remains above, works the dung up into elliptical pellets and lowers them down the shaft, while the female, after laying an egg in the sand at the bottom of the burrow, receives the pellets, tears them apart and packs the fragments down, as if she were working in a silo, till they form a mass as big as one's finger. Then she digs in succession a few branch galleries off from the main shaft, furnishes each of them with an egg and provisions it in the same manner. By constructing an ingenious apparatus and providing the beetles with an unlimited supply of manure, Fabre induced one male to make 239 pellets and hand them down to the female, but unfortunately the latter had died at the bottom of the gallery, so that there were no

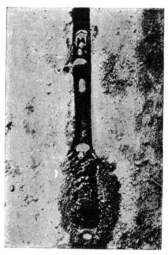

Fig. 5. — Lowermost portion of burrow of the Minotaur (*Minotaurus typhœus*) showing the male beetle lowering the dung in pellets and the female storing it in layers above her egg. (After J. H. Fabre.)

eggs and the pellets had not been torn apart and stored. The development of the young requires five months and the female very probably remains in the burrow till the brood emerges and crawls up to the surface.

2. *Passalidœ* — These large, active, jet-black, flattened and parallel-sided beetles (Fig. 6) are common throughout the tropics of both hemispheres. A single species, *Passalus*

cornutus, occurs in the United States as far north as Michigan and Massachusetts. Ohaus,[15] who first studied the habits of several species in the forests of Brazil, has shown that they form colonies consisting of a male and female and their progeny, and make large, rough galleries in rather damp, rotten logs. The broadly elliptical yellowish green or greenish black eggs, to the number of a dozen or more, are laid in a loose cluster and guarded by the parents. The

Fig. 6. — *Passalus sp.* Adult beetle and rather young larva, about twice natural size.

larvæ are drab-colored and cylindrical, with the hind pair of legs reduced to peculiar short paw-like appendages which can be rubbed back and forth on striated plates at the bases of the middle legs (Fig. 7), thus producing a shrill note. On the dorsal surface of the abdomen of the adult beetle there is also a stridulatory organ in the form of patches of minute denticles which may be rubbed against similar structures on the lower surfaces of the wings.[16]

Ohaus found that the parent beetles triturate the rotten wood and apparently treat it with some digestive secretion which makes it a proper food for the larvæ, since their mouth-parts are too feebly developed to enable them to attack the wood directly. They are therefore compelled to follow along after their tunnelling parents and pick up the prepared food. All the members of the colony are kept

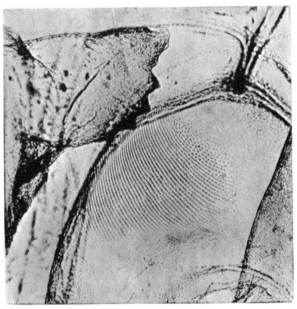

Fig. 7. — Microphotograph showing abbreviated, paw-like hind leg of Passalus larva and the striated surface over which its toothed edge is rubbed during stridulation.

together by stridulatory signals. The noise made by the beetles is so loud that it is possible to detect the presence of a Passalus colony in a log by merely giving it a few sharp raps. I have been startled on more than one occasion in Central America by the shrill response thus elicited from large Passali that were burrowing deep in the wood. When the larvæ are mature they pupate in the burrows and the emerging beetles are guarded and fed by the parents till they are fully mature. Observations that I have made in Australia, Central America, Trinidad and British Guiana confirm Ohaus's statements.

3. *Phrenapates* — Nearly a century ago Kirby [17] described a peculiar beetle from Colombia as *Phrenapates bennetti* (Fig. 8). It is about an inch long, jet-black and

FIG. 8. — *Phrenapates bennetti,* a social Tenebrionid beetle, from a specimen in the Museum of Comparative Zoology. (Photograph by Mr. Leland H. Taylor.)

shining and superficially resembles Passalus, but belongs to a very different family, the Tenebrionidæ. G. C. Champion [18] records it from Central America (Panama to Guatemala), and states that he " met it in plenty in decaying timber in the humid forest region of Chiriqui and frequently dug it out of cylindrical burrows, probably made by the larvæ, in the solid wood." Some years later Ohaus encountered the insect in Ecuador and gave a more detailed account of its habits.[20]   The male and female gnaw in the wood of the silk-cotton tree (Bombax) a narrow, cylindrical gallery about a foot and a half long and make roomy niches on each side of it at definite intervals.   All the work is neat and smooth, unlike the burrow of Passalus.   In each of the niches Ohaus found an egg or one or two larvæ, the latter feeding on fine, elongate shavings which filled the

Fig. 9. — Tachigalia beetles, the larger *Coccidotrophus socialis*, the smaller *Eunausibius wheeleri*.

niches and had evidently been provided by the parent beetles.   The eggs are laid at rather long intervals so that the larvæ, unlike those of the Passalus, vary considerably in size.   They resemble our common meal-worms (*Tenebrio molitor*), but are milk-white.   No stridulatory organs could be detected in the beetles, but like some other Tenebrionids (Blaps) they emit a penetrating odor.

4. *Tachigalia Beetles* — During the summer of 1920 I discovered in the jungles of British Guiana a couple of Silvanid beetles which lead a more spectacular existence than some of the preceding.[21]   These beetles, which Messrs. Schwarz and Barber [22] have named *Coccidotrophus socialis* and *Eunausibius wheeleri* (Fig. 9), are less than a quarter

of an inch in length and have long, slender, sub-cylindrical, red or chestnut brown bodies, with short legs and club-shaped antennæ. They occur only in the hollow leaf-petioles (Fig. 10) of a very interesting tree, *Tachigalia panicu-*

FIG. 10.—Bases of leaf-petiole of *Tachigalia paniculata*. (a) of young, shade tree; (b) of large, sun tree, both nearly one-half natural size. Pieces of the older petiole and adjacent trunk have been cut out to show the cavity.

lata, and only in young specimens 1½ to 7 ft. high while they are growing in the shade under the higher trees of the jungle. The older trees, which may attain a height of 40 feet or more, have all their petioles inhabited by viciously stinging or biting ants. Each beetle colony is

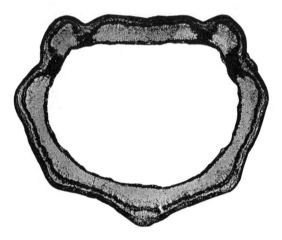

Fig. 11. — Cross-section of base of a young, uninhabited petiole of *Tachigalia paniculata* showing the bands of protein-containing nutritive tissue (dark). (Photograph by Professor I. W. Bailey.)

started by a male and female which bore through the wall of the petiole, clean out any pith or remains of previous occupants it may contain and commence feeding on a peculiar tissue rich in proteins, which is developed in parallel, longitudinal strands in the wall of the petiole (Figs. 11 and 12). As they keep gnawing out this tissue they gradually make grooves and pile their feces on the ungnawed intervening areas, so that the interior of the petiole assumes a peculiar appearance (Figs. 13 and 14). While the beetles are thus engaged numbers of small mealy-bugs of the genus *Pseudococcus* (*Ps. bromeliæ*) (Fig. 15), covered with snow-white wax, wander into the petiole through the opening made by the beetles, settle in the grooves, sink their delicate sucking mouth-parts into the nutritive tissue and imbibe its juices. The beetles soon begin to lay their small, elliptical, white eggs along the edges of the grooves (Fig. 14), and the hatching larvæ, which are beautifully translucent, run about in the cavity and feed on the same tissue as the parents. But incredible as it may seem both the adult beetles and the larvæ in all stages have learned to stroke the mealy-bugs with their antennæ, just as our common ants stroke similar mealy-bugs and plant-lice, and feed on the droplets of honey-dew, or saccharine excrement which they give off when their backs are properly

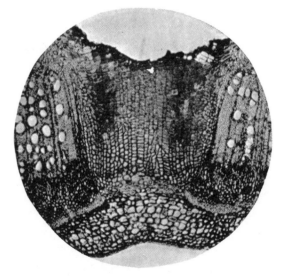

Fig. 12. — Enlargement of one of the bands of nutritive tissue of the preceding figure, showing the rather homogeneous protein-containing cells. (Microphotograph by Professor I. W. Bailey.)

titillated. So greedy are the Silvanids for this nectar that I have seen a beetle or a larva stroke a mealy-bug for an hour or longer and receive and swallow a drink every few

minutes. When two or more beetles or two or more larvæ
or a group of beetles and larvæ happen to be engaged in
stroking the same mealy-bug, they stand around it, like so
many pigs around a trough, and the larger or stronger individual keeps butting the others away with its head. The
butted individuals, however, keep returning and resuming
their stroking till the knocks become too severe or the

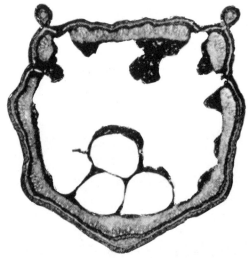

FIG. 13. — Cross-section of Tachigalia petiole inhabited by a
flourishing colony of *Coccidotrophus socialis*. The gnawed out
areas of nutritive tissue are seen above, with the frass piled on
the intermediate areas; below three cocoons have been sectioned.
(Photograph by Professor I. W. Bailey.)

stronger individual leaves and begins to stroke another
mealy-bug. Thus the beetles and their progeny have discovered a rich food supply, consisting in part of the pro-
teid-containing tissues of the Tachigalia and in part of the
sugar and water discharged by the mealy-bugs, which in
turn imbibe the sap of the tree. The beetles lay their eggs
at intervals so that larvæ in all stages are found in the same

colony. When mature each larva constructs a cocoon of minute particles bitten out of the plant tissues (Fig. 16), creeps into it, closes the opening from the inside and pupates.[23] When the young beetles hatch they remain with their parents and soon begin to lay eggs, so that eventually

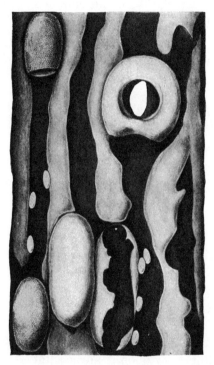

Fig. 14. — Enlarged drawing of a part of the wall of a *Tachigalia* petiole inhabited by *Coccidotrophus socialis;* showing the food grooves and frass ridges, the entrance with its wall, the eggs, an intact and broken cocoon of the *Coccidotrophus* and two cocoons of the Coccid parasite, *Blepyrus tachigaliœ,* one of them after the eclosion of the parasite.

the colony consists of several dozen beetles, larvæ, pupæ, and mealy-bugs in all stages and all living peacefully together, except for the little family bickerings of the

beetles and larvæ over the milking of their patient, snow-
white cattle. When the petiole becomes too crowded, pairs
of young beetles leave it, enter other petioles of the same
or other Tachigalia trees and start new colonies. As the
tree grows and emerges from the undergrowth into the sunlight, the ants which then take complete possession of it
oust the beetles from the petiolar cavities but adopt their
mealy-bugs, just as the invading German army appropriated the French cattle. There are many other extraordi-

Fig. 15. — *Pseudoccus bromeliæ* Bouché. Sketch of an adult
living female with intact covering and peripheral pencils of wax.

nary insects associated with the Tachigalia, its beetles and
mealy-bugs, but I must omit an account of them because
they are irrelevant to the present discussion.

5. *Ipid Ambrosia Beetles* — The family Ipidæ comprises
small, cylindrical, red-brown or black beetles which
live in the trunks and branches of trees.[24] The group is now
divided into two sections, one of which includes the bark-
beetles, which are nonsocial and make the beautiful, radi-
ating burrows so commonly seen on the inner surface of the
bark of sickly trees, the other includes the ambrosia beetles

(Fig. 17), which are social and run their burrows right into the wood of healthy or recently felled trees. The name "ambrosia beetles" is derived from a term applied by Schmidberger to the fungi which the beetles cultivate as food for themselves and their larvæ. Structurally the two sections of the family Ipidæ can be readily distinguished by the mouthparts, the bark-beetles having their maxillæ armed with a row of 12 to 20 strong tooth-like bristles

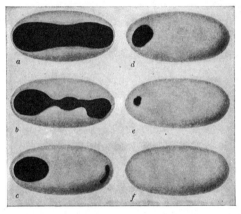

Fig. 16.— Six successive stages in the construction of the cocoon by the full-grown larva of *Coccidotrophus socialis.*

adapted to gnawing bark, whereas the maxillæ of the ambrosia beetles are fringed with 30 to 40 delicate, curved bristles, evidently suited to cropping the soft hyphæ of their food-fungus. Fourteen genera and nearly four hundred species of ambrosia beetles have been described. One genus alone, Xyleborus, which is cosmopolitan, contains 246 species. The fungi that grow in the galleries often give their walls a black stain, so that the value of the wood thus affected is greatly impaired. One species, *Xyleborus perforans,* has a bad reputation in the tropics, where it goes by the name of "tippling Tommy," because it has a strong predilection for boring in the staves of wine, beer and rum casks and thus causing much leakage. It might be adopted by our prohibitionists as their totem-animal.

The ambrosia beetles were first carefully studied in this country by H. G. Hubbard, whose untimely death deprived

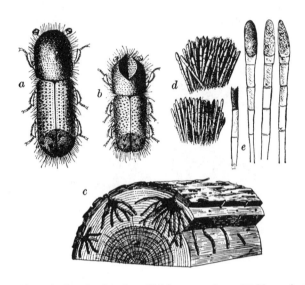

Fig. 17. — Ambrosia beetle, *Xyleborus celsus* Eichh., of the hickory (after H. G. Hubbard); *a,* female beetle; *b,* male; *c,* piece of hickory, showing burrows of *X. celsus* in the sap-wood; *d,* ambrosia grown by *X. celsus* on the walls of the burrows; *e,* same more enlarged.

us of one of our most talented entomologists.[25]   I can not do better than quote his concise account of two of our species of Pterocyclon (*mali* and *fasciatum*): " The sexes are alike, and the males assist the females in forming new colonies.   The young are raised in separate pits or cradles which they never leave until they reach the adult stage. The galleries, constructed by the mature female beetles, extend rather deeply into the wood, with their branches mostly in a horizontal plane. The mother beetle deposits her eggs singly in circular pits which she excavates in the gallery in two opposite series, parallel with the grain of the wood. The eggs are loosely packed in the pits with chips and material taken from the fungus bed which she has previously prepared in the vicinity and upon which the ambrosia has begun to grow. The young larvæ, as soon as they hatch out, eat the fungus from these chips and eject the refuse from their cradles. At first they lie curled up in the pit made by the mother, but as they grow larger, with their own jaws they deepen their cradles, until, at full growth, they slightly exceed the length of the larvæ when fully extended. The larvæ swallow the wood which they excavate, but do not digest it. It passes through the intestines unchanged in cellular texture, but cemented by the excrement into pellets and stained a yellowish color. The pellets of excrement are not allowed by the larvæ to accumulate in their cradles, but are frequently ejected by them and are removed and cast out of the mouth of the borings by the mother beetles. A portion of the excrement is evidently utilized to form the fungus bed. The mother beetle is constantly in attendance upon her young during the period of their development, and guards them with jealous care. The mouth of each cradle is closed with a plug of the food fungus, and as fast as this is consumed it is renewed with fresh material. The larvæ from time to time perforate this plug and clean out their cells, pushing out the pellets of excrement through the opening. This debris is promptly removed by the mother and the opening again sealed with ambrosia. The young transform to perfect beetles before leaving their cradles and emerging into the galleries." The ambrosia of Pterocyclon " is moniliform and resembles a mass of pearly beads. In its incipient stages a formative stem is seen, which has short joints that become globular conidia and break apart. Short chains of cells, sometimes showing branches, may often be separated from the mass. The base of the fungous mass is stained with a tinge of green, but the stain of the wood is almost black.

6. *Platypodid Ambrosia Beetles* — These were formerly included among the Ipidæ but are now regarded as an independent family.[26] They can be easily distinguished by their much broader head and longer feet, the first joint of the tarsi being as long as all the remaining joints together. The great majority of the species are tropical, so that their habits have not as yet been very thoroughly studied. So far as known, the Platypodids all bore in the wood of dying or recently felled trees, live in societies and feed on fungi which they grow on the walls of their burrows. Hubbard and Swaine have studied some of our North American and Strohmeyer has published some observations on one of the few European forms.[27] The following description of *Platypus compositus* is quoted from Hubbard: "They are powerful excavators, generally selecting the trunks of large trees and driving their galleries deep into the heart-wood. They do not attack healthy trees but are attracted only by the fermenting of the sap of dying or very badly injured trees. The death rattle is not more ominous of dissolution in animals than the presence of these beetles in standing timber. . . . The female is frequently accompanied by several males and as they are savage fighters, fierce sexual contests take place, as a result of which the galleries are often strewn with fragments of the vanquished. The projecting spines at the ends of the wing-cases are very effective weapons in these fights. With their aid a beetle attacked in the rear can make a good defense and fre-

quently by a lucky strike is able to dislocate the out-stretched neck of his enemy. The females produce from 100 to 200 elongate-oval pearl-white eggs, which they deposit, in clusters of 10 or 12, loosely in the galleries. The young require five or six weeks for their development. They wander about in the passages and feed in company upon the ambrosia which grows here and there upon the walls. . . . The older larvæ assist in excavating the galleries, but they do not eat or swallow the wood. The larvæ of all stages are surprisingly alert, active and intelligent. They exhibit curiosity equally with the adults, and show evident regard for the eggs and very tender young, which are scattered at random about the passages, and might easily be destroyed by them in their movements. If thrown into a panic the young scurry away with an undulatory movement of their bodies, but the older larvæ will fre-quently stop at the nearest intersecting passage and show fight to cover their retreat." The ambrosia of *P. com-positus* consists of hemispherical conidia growing in clusters on branching stems. The long-continued growth of this fungus blackens the walls of the older galleries.

Each species of ambrosia beetle — and this is true of both the Ipidæ and the Platypodidæ — grows its own peculiar fungus in a pure culture, irrespective of the tree it may select for its burrows. Strohmeyer seems to have shown how in the case of certain Platypodids the mother beetle manages to obtain the spores of the particular fungus which she cultivates.[28] He finds that she carries them from the burrows in which she passed her larval and pupal stages to the new burrows which she makes for her own progeny in a kind of crate or basket consisting of one or several dense tufts of long, curved hairs on the top of her head or on her mouth-parts; and Schneider-Orelli has found that the females of the Ipid ambrosia beetles carry the fungus in the fore part of the stomach and are thus able to infect the walls of the new burrows which they establish.[29] These are only two of the instances among the social insects of the actual transmission of a food-plant from generation to generation.

We may now summarize very briefly the main points of interest in connection with the social beetles:

1. The six unrelated families are all very ancient. Species of four of them (Silvanidæ, Tenebrionidæ, Ipidæ and Platypodidæ) are, in fact, known from the Baltic Amber. The absence of the dung-beetles from that formation is easily explained, since these insects are not arboreal, nor are they attracted by liquid resins. Several of the living genera (Scarabæus, Copris, Onthophagus, Sisyphus, and Gymnopleurus), however, are known from the Upper Miocene shales of Oeningen, and Hagedorn mentions several species of ambrosia beetles as occurring in the African and Malagasy copal, a fossil resin of comparatively recent formation. There can be little doubt that all the six families which I have been considering are much older than these records would seem to indicate. Most of them, in fact, are cited by Handlirsch as probably having arisen at the beginning of the Cretaceous or even earlier.[30]

2. The substances on which the six groups of social beetles feed are remarkably diverse, ranging from dung and wood in various stages of decay to the living tissues of plants, the honey-dew of mealy-bugs and delicate fungi. These are all abundant and ubiquitous substances of vegetable origin, and all the social beetles manage to store or find their food in such peculiar places that they can avoid intense competition with most other organisms.

3. This abundant but in many cases not very nutritious

food-supply which the adult beetles seek and exploit primarily for their own consumption enables them to acquire a considerable longevity, and this in turn, of course, enables them to survive the hatching and development of their young.

4. In all the groups the parent beetles show a very pronounced interest in their offspring, and feed them directly or, at any rate, place them in close contact with the food and guard them.

5. The father beetle cooperates to a greater or less extent with the mother beetle in providing for the young, although his cooperation may be slight. Probably it is really *nil* in most of the Ipid ambrosia beetles, the males of which are in many species wingless and very rare, so that mating must take place in the maternal colony.

6. There are neither structural nor physiological differences between the fully developed young and the adult parents of the social beetles. In other words, nothing like a development of castes has made its appearance among them.

LECTURE II

WASPS SOLITARY AND SOCIAL

In the preceding lecture I gave a brief account of the rudimentary social life of certain beetles and called attention to the fact that in all or nearly all of them the male cooperates with the female parent in victualing or protecting the offspring. I endeavored to show that all these societies have their inception or *raison d'être* in the specialized feeding habits of the parents and that in all of them the food is of vegetable origin, abundant but not very nutritious in some of the cases (dung and rotten wood in the Scarabæidæ, Passalidæ and Phrenapates), in others highly nutritious, but obtainable only in small quantities at a time (living plant-tissues and honey-dew in the case of the Tachigalia beetles, ambrosia of the Ipidæ and Platypodidæ). The adequate exploitation of such food-supplies is necessarily time-consuming and has evidently led to a lengthening of the adult lives of the beetles. This in turn has naturally brought about an overlapping of the juvenile by the parent generation, thus enabling the parents to acquire contact and acquaintance with their young and an interest in providing them with the same kind of food as that on which they themselves habitually feed. In the insects which I shall consider in this lecture, we find a series of societies originating in a very different type of feeding and leading to much more complicated and more definitely integrated associations.[1]

Although the wasps have attracted fewer investigators than the ants and bees, they are of even greater interest to the student who is tracing the evolution of specialized instincts and social habits. The wasp group is one of enormous size and is really made up of two great complexes, the Sphecoids and the Vespoids, together comprising more than a dozen families and some 10,000 species. Of these only about 800 are clearly social. We have more or less fragmentary behavioristic studies of scarcely 5 per cent. of all the species. Yet they cover a sufficient number of forms to enable us to establish the following generalizations:

1. The structure and behavior of the Sphecoids and Vespoids show that they must have arisen from what have been called Parasitic Hymenoptera, and the structure of the ants and bees shows that they in turn must have arisen from primitive Sphecoids or Vespoids.

2. The social wasps comprise several groups which have evolved independently from primitive, solitary Vespoids. There are also a few Sphecoids that exhibit subsocial propensities.

3. Both the Sphecoids and the Vespoids are primarily predaceous and feed upon freshly captured insects, but the adults are fond of visiting flowers and feeding on nectar. Some social wasps store honey in their nests, but it is probably not an exclusive or essential constituent of the larval food. One small and aberrant group of solitary Vespoids, the Masarinæ, however, provision their cells with a paste of honey and pollen, like the solitary bees. The insect prey on which at least the young of nearly all the wasps subsist, being rich in fats and proteids, is an ideal food, though it has to be provided in larger quantity than such concentrated vegetable substances as pollen and nectar. It is also scarcer and more difficult to obtain. Hence the

definite tendency in adult wasps towards a honey regimen at least for the purpose of eking out the primitive animal diet.

4. We are able to observe in the social wasps more clearly than in other social insects the peculiar phenomenon which I have called "trophallaxis," *i.e.,* the mutual exchange of food between adults and their larval young.

5. The study of the wasps and of their ancestors among the Parasitic Hymenoptera furnishes us with a key to the understanding of the parthenogenesis and the peculiar dominance of the female sex (gynarchy), which is retained throughout the whole group of stinging Hymenoptera (wasps, bees and ants).

6. In the social wasps we witness the first gradual development of a worker caste and also of polygyny and swarming.

7. We observe in wasps a high degree of modifiability of behavior and an extraordinary development of memory, endowments which have led McDougall to claim for them " a degree of intelligence which (with the doubtful exception of the higher mammals) approaches most nearly to the human," and Bergson to point to their activities as one of the most telling arguments in favor of his intuitional theory of instinct.[2] Although I believe that these and many other authors have been guilty of some exaggeration, the wasp's psychic powers compared with those of most other insects or even of many of the lower Vertebrates seem to me, nevertheless, to be sufficiently remarkable.

We shall have to examine each of these generalizations more closely. Some of them may be considered forthwith, others more advantageously after the description and illustration of a selected series of species.

Recent studies of the parasitic, or as I prefer to call them with O. M. Reuter, the "parasitoid" Hymenoptera,[3] have revealed certain peculiar traits which recur in a modified form in the behavior of their Sphecoid and Vespoid descendants. But what are these parasitoids? You are all familiar with the fact that a large number of insects regularly lay their eggs on or in plants and that the hatching larvæ devour the plant tissues and eventually pupate and emerge as insects which repeat the same cycle of behavior. There is, however, another immense, but less conspicuous, assemblage of insects that lay their eggs on or in the living eggs, larvæ, pupæ and adults of other insects, and the eggs thus deposited develop into larvæ which gradually devour the softer tissues in which they happen to find themselves. Species that behave in this manner are not true parasites, but extremely economical predators, because they eventually kill their victims, but before doing so spare them as much as possible in order that they may continue to feed and grow and thus yield fresh nutriment just as it is needed. For this reason and also because, as a rule, only the larval insect behaves in the manner described, it is best called a "parasitoid." The adult into which it develops is, in fact, a very highly organized, active, free-living creature, totally devoid of any of the stigmata of "degeneration" so common among parasites, and with such exquisitely perfected sensory, nervous and muscular organs that it can detect its prey in the most intricate environment and under the subtlest disguises.

The parasitoids exhibit another peculiarity which was destined to acquire great importance in their descendants, the wasps, bees and ants, namely parthenogenesis, or the ability of the female to lay unfertilized eggs capable of complete development. As a rule, if not always, these parthenogenetic eggs develop into males, whereas fertilized eggs laid by the same female develop into individuals of her own sex. Thus the female has become to some extent independent of the male in the matter of reproduction. It will be seen that if the parthenogenetic egg were able to develop into a female, as it frequently does in certain insects like the plant-lice, the male might become entirely superfluous. There are a few insects in which this has occurred or in which the male appears only at infrequent intervals in a long series of generations. But matters have not come to such a pass in the parasitoids or in the wasps, bees and ants, though these insects have perfected another method of reducing the male to a mere episode in the life of the female. Individuals of this sex are provided with a small muscular sac, the spermatheca, which is filled with sperm during the single act of mating, and this sac is provided with glands, the secretion of which may keep the sperm alive for months or even years. According to a generally accepted theory, the female can voluntarily contract the wall of her spermatheca and thus permit sperm to leave it and fertilize the eggs as they are passing its orifice on their way to being laid, or she can keep the orifice closed and thus lay unfertilized eggs. The mother can thus control the sex of her offspring or if she has failed to mate, or exhausted all the sperm in her spermatheca, may nevertheless be able to lay male-producing eggs. There seems also to be something compensatory, or regulatory, in this ability of the female parasitoid to produce males parthenogenetically, for if she be unable to meet with a male — and this predicament is very apt to arise among such small and widely dispersed animals as insects — she can produce the missing sex and thus increase the chances of mating for the next generation of females.

Certain facts indicate that the sex of the egg may not be

determined in the manner here described, but their consideration must be postponed till they can be taken up in connection with the honey-bee. We are justified, notwithstanding, in regarding the female parasitoid, wasp, bee or ant, after she has appropriated and stored in her spermatheca all the essential elements of the male, as a potential hermaphrodite. The body, or soma, of the male, after mating, thus really becomes superfluous and soon perishes. In the solitary wasps the male is a nonentity, although in a few species he may hang around and try to guard the nest. But in the bees, ants and social wasps he has not even the status of a loafing policeman, and all the activities of the community are carried on by the females, and mostly by widows, debutantes and spinsters. The facts certainly compel even those who, like myself, are neither feminists nor vegetarians, to confess that the whole trend of evolution in the most interesting of social insects is towards an ever increasing matriarchy, or gynarchy and vegetarianism.

Now if we carefully observe a parasitoid while she is ovipositing in her prey, we obtain a clue to the meaning of the peculiar behavior of the solitary wasps which has led Fabre to certain erroneous conclusions and philosophers like Bergson to his peculiar interpretation of instinct. The parasitoid is furnished at the posterior end of her body with a well-developed ovipositor, a slender, pointed instrument for piercing the tough integument of her victim. But this instrument also has another function, namely, that of making punctures through which droplets of the victim's blood may exude and be devoured by the parasitoid. She may often be seen thrusting her ovipositor into her prey without ovipositing and merely for the sake of obtaining food, or she may feed at a puncture she has made while ovipositing.[4] Obviously feeding and oviposition are here congenitally, or hereditarily conditioned reflexes, to use Pawlow's expression. In other words, the internal hunger and reproductive stimuli, or appetites, are so intimately associated with one another that mere contact with the

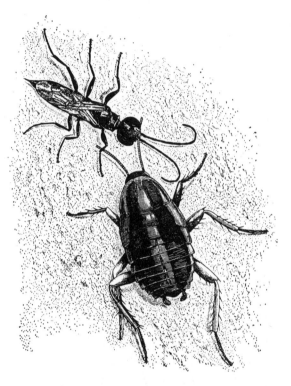

Fig. 18. — *Dolichurus stantoni* of the Philippines dragging a young cockroach (*Blatella bisignata*) to her burrow. x 6. (After F. X. Williams.)

prey releases either the feeding or the ovipositing reactions, or both. And, of course, both of these reactions are purely selfish, the one being concerned with getting food, the other with getting relief from the discomfort of egg-pressure in the ovaries, and both may initiate elaborate trains or pat-

terns of behavior (instincts).[5] This is true not only of the
parasitoids but also of insects in general.

Turning now to the solitary wasps we find that, like the
parasitoids, they prey on other insects and that each species
of wasp usually has a predilection for a particular species
(Figs. 18 and 19), genus or family of insects, or even for
a particular sex, as in the case of one of our common wasps,
Aphilanthops frigidus, which preys only on queen ants.[6]
The chief difference betwen the parasitoid and the solitary
wasp lies in the fact that the latter lays her egg on or near
her victim after stinging it till it is motionless. The sting
is merely the ovipositor which is now used only for defence
or for reducing the prey to impotence, while the mouth-
parts and especially the mandibles are used for obtaining
food. Many solitary wasps, after stinging their prey, devour it in part or entirely, or chew, *i.e.,* malaxate, its neck
and lap up the exuding juices. This behavior is essentially
like that of the parasitoid, and in its more frequent, feebler
manifestations may be regarded as a vestigial feeding. The
adult wasp is no longer as carnivorous as its ancestors, because she has come to rely to some extent on the energizing nectar of flowers, but this substance contains no proteins and is therefore an improper food for her growing
larval young. Roubaud and Rabaud have recently shown
that the stinging of the prey follows reflexly as soon as it
has been seized and comes in contact with the wasp's
sternum, and that the accidental position of the prey when
it thus releases the reflex determines the point where it
will be stung. Moreover, the stinging is repeated till the
victim ceases to struggle and becomes motionless. Hence
the stinging does not occur in the schematic manner nor
necessarily in the nerve ganglia, as described by Fabre.
It has also been shown that the venom introduced into the

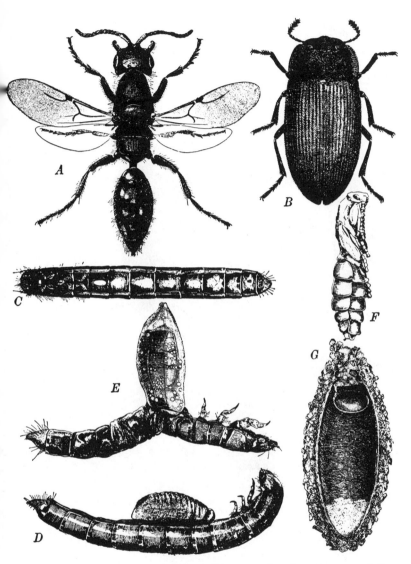

Fig. 19.—*A*, Female of a Bethylid wasp, *Epyris extraneus*, of the
Philippines; *B*, Tenebrionid beetle, *Gonocephalum seriatum*; *C*, Larva
of the same with egg of *E. extraneus* on middle of ventral surface; *D*,
Young *E. extraneus* larva feeding on the larva of *G. seriatum*; *E*, Later
stage of same; *F*, Pupa of *E. extraneus*; *G*, Cocoon of same. (After
F. X. Williams.)

tissues of the prey by the sting produces paralysis or even
death and also acts as an antiseptic in preserving the prey
from decomposition for weeks or even months while the
larva that hatches from the wasp's egg is feeding on the tissue, but these properties of the venom are accidental and
unforeseen. Hence Fabre's and Bergson's contention that
the solitary wasp is a clairvoyant surgeon, with an intuitive

Fig. 20. — *Sphex procerus* carrying caterpillar of sphinx moth to her
burrow. (Photograph by Prof. Carl Hartman.)

knowledge of the internal anatomy of the particular insect
on which it preys, may be dismissed as a myth.

The explanations here given of the malaxation and stinging of the prey are purely physiological, but it is not at all
certain that such explanations are applicable to the entire
behavior cycle of the solitary wasp. Before inquiring into
this matter, it will be advisable to sketch very briefly the behavior of a typical Sphex as a paradigm of the whole group

of Sphecoids and solitary Vespoids. The female Sphex, after mating, digs in sandy soil a slanting or perpendicular tunnel and widens its end to form an elliptical chamber. She may thereupon close the entrance, rise into the air and fly in undulating spirals over the burrow, thus making what is called a " flight of orientation," or " locality study," because it enables her to fix in her sensorium the precise

Fig. 21.—*Sphex procerus* carrying chips of wood to throw into the burrow at the left of the figure. (Photograph by Prof. Carl Hartman.)

position of the burrow in relation to the surrounding objects, so that she may find the spot again. Then she flies off in search of her prey, which is a particular species of hairless caterpillar (Fig. 20). When it is found, she stings it into insensibility, malaxates its neck, while imbibing the exuding juices, and drags it or flies with it to the entrance of her burrow. Here she drops her victim and, after entering and inspecting the burrow, returns and takes it down into the chamber, glues her egg to its surface and closes the burrow by filling it with sand or detritus collected from the surrounding soil (Figs. 21 and 22). As soon as the next egg matures in her ovaries she proceeds to repeat the same behavior cycle at some other spot. In the meantime the provisioned egg hatches, and the larva, after devouring the

Fig. 22. — Burrow of *Sphex procerus* in section, showing filling of débris in the tunnel and the paralyzed sphinx moth caterpillar in the cell, with the egg glued to its side. (Photograph by Prof. Carl Hartman.)

helpless caterpillar, spins a cocoon, pupates *in situ* and eventually emerges as a perfect Sphex.

Some of our species of Sphex actually tamp down the filling of their burrows with a small, carefully selected pebble, held in the mandibles and used as a hammer or

pestle (Fig. 23). This astonishing behavior, which has
been carefully observed by no less than nine investigators
(Williston, Pergande, G. W. and E. G. Peckham, Hartman, Hungerford and Williams, and Phil and Nellie Rau)
can hardly be reduced to simple physiological reflexes. The
same would seem to be true of the orientation flight and
return to the burrow and the fact that some species of
Sphex provide the egg with a single large caterpillar, others
with several small caterpillars, but in all cases with just

Fig. 23. — *Sphex urnarius* using a selected pebble to pound down
earth over burrow. (After G. W. and E. G. Peckham.)

enough food to enable the larva to grow to the full stature
of a normal individual of its species. The question also
arises as to the proper interpretation of the peculiar predilection of the wasp for a particular species of prey. This
seems to be the more inexplicable, because experiment has
shown that the larva can be successfully reared when some
very different insect is substituted for the species which it
habitually devours. As I am emphasizing the rôle of nutrition in these lectures, I shall digress somewhat on this
question of food specialization, and in order to bring the

matter before you as vividly as possible recast the behavior of Sphex in the form of a tragic drama in three acts, with the following brief synopsis:

Act I. A sandy country with sparse vegetation inhabited by caterpillars and other insects. Time, a hot, sunny day in early August. Scene 1. Miss Sphex arrayed in all the charm of maidenhood being courted by Mr. Sphex. Wedding among the flowers. Scene 2. Mrs. Sphex, deserted by her scatter-brained spouse, settles down and excavates a kind of cyclone-cellar. She closes its door and leaves the stage.

Act II. Scene 1. Same as in Act I. Mrs. Sphex, hunting in the vegetation, finds a caterpillar, struggles with it, stings it and gnaws its neck till it lies motionless. Scene 2. She drags it into the cellar and placing her offspring on it behind the scenes, returns and at once leaves the stage after locking the door, amid a storm of applause.

Act III. Scene 1. Interior of Mrs. Sphex's cellar. Baby Sphex slowly devouring caterpillar till only its skin remains. Scene 2. Baby Sphex, now a large, buxom lass, weaves an elaborate nightgown for herself and goes to bed as the curtain falls.

As a work of art this drama is defective, because the climax, the stinging of the caterpillar, falls in the early part of the second act, and because the heroine leaves the stage soon afterwards for good, as if she had been suddenly taken ill and had to substitute her drowsy offspring to perform the whole third act. Still this is the sequence in which the drama is related by all the observers, and I have presented my account in the same manner, because it has undeniable advantages. But see what happens when we rearrange the drama by making the third or last act the first, and the first and second the second and third, respectively. There is then only one heroine who holds the center of the stage throughout the performance. We witness her gradual growth and development from infancy during the first act, her wedding, desertion and cellar-excavating exploits during the second, and the thrilling chase, stinging and entombment of the hereditary victim in the third act.

I have just committed the unpardonable sin of humanizing the wasp, but being desirous of making my point perfectly clear, I am going to do something still more scandalous and ask you for a moment to vespize the human being. Suppose that the human mother were in the habit of carefully tying her new-born baby to the arm-pit of a paralyzed elephant which she had locked in a huge cellar. The baby — we must, of course, suppose that it is a girl baby — is armless, legless and blind, but has been born with powerful jaws and teeth and an insatiable appetite. Under the circumstances she would have to eat the elephant or die. Supposing now that she fed on the elephant day after day between naps till only its tough hide and hard skeleton were left, and that she then took an unusually long nap and awoke as a magnificent, winged, strong-limbed amazon, with a marvelously keen sense of smell and superb eyes, clad in burnished armor and with a poisoned lance in her hand. With such attractions and equipment we could hardly expect her to stay long in a cellar. She would at once break through the soil into the daylight. Now suppose she happened to emerge, with a great and natural appetite, in a zoological garden, should we be astonished to see her make straight for the elephant house? Why, she would recognize the faintest odor of elephant borne to her on the breeze. She would herself be, in a sense, merely a metabolized elephant. Of course, we should be startled to see her leap on the elephant's back, plunge her lance into its arm-pit, drag it several miles over the ground, hide it in a cellar and tie her offspring to its hide.

The point I wish to make is this: We have all along in our accounts treated the life-history of the insect as that of two individuals in such a manner as to obscure or obliterate the experience of the individual. We begin with the full-fledged insect descending from the blue, and then describe her behavior as if it were a pure inheritance or improvisation. But when we describe her activities as those of a single individual from the beginning of her development to death, we find that the adult female, before she begins to make and provision her nest, has probably learned something from her long and intimate larval contact with the environment. For months she has inhabited a chamber like the one she will excavate or build for her own progeny, for days she has been devouring a particular species of caterpillar, and she has even dug a sufficient distance through the soil to be familiar with its properties. She possesses, therefore, a certain amount of acquaintance with soil and with caterpillars. That this should persist as memory is not only possible but extremely probable when we consider that the central nervous system of the larva passes without profound change into that of the adult wasp, and that the latter shows unmistakable evidence of possessing a remarkable memory when she makes such locality studies as have been described and returns to her nest or prey after an absence of several hours or even days. We are also enabled to understand why the wasp confines her attention to a particular species or even to one sex of a species while searching for her prey, and why the malaxation or mutilation of the prey may be regarded as a reminiscent act of feeding. In brief, all those activities of the adult wasp which are partly or wholly interpretable as a repetition of larval behavior, may be attributed to memory — not in the sense of recollection, with its feeling of "pastness," but of mere sensory and motor memory, the *memoria sensitiva* of scholastic writers, or the "associative memory" of comparative psychologists.

But there still remain unexplained the more striking activities, those performed for the first time in the wasp's life history, namely the cocoon-spinning of the larva, the making, closing and opening of the nest, oviposition, etc. No doubt these acts are all initiated by stimuli, partly internal and partly external, such as hunger, the tension of accumulated silk in the spinning glands, of eggs in the ovaries, hormones in the blood, and olfactory and tactile impressions from contact with the caterpillar and the soil, but the reeling off of the train of these purposive responses must depend on inherited dispositions in some way correlated with the structure of the nervous and muscular apparatus. And we must suppose that these dispositions somehow represent the experience of untold former generations of wasps. We are, however, unable to form any adequate conception of the extent of the racial experience of the solitary wasps as a group, and therefore of the amount of condensation or syncopation with which it is epitomized in the behavior of the individual wasp, and this disability on our part is largely responsible not only for the old supernatural conceptions of instinct but also for theories like those of Bergson, the Neodarwinians and the mutationists.

We may now turn to the evolution of social behavior, which, in diverging lines of descent, has been gradually evolved and perfected from such a method as that employed by most of the Sphecoids and non-social Vespoids. This method, which consists in rapidly accumulating an

amount of prey sufficient to enable the young to develop
to maturity, and of then closing the cell before the egg has
hatched, we may designate, with Roubaud, as "mass provisioning." We have seen that in some cases the mother
wasp stores a single large insect, in others a number of
smaller ones, before closing the cell. If in the latter case
the accumulation of the prey is delayed on account of scarcity or inclement weather, the egg, which has been glued
to the first small insect captured, hatches before the mother
wasp has succeeded in collecting a sufficient supply for the
growth of the larva. She is therefore reduced to feeding her
offspring from day to day, *i.e.,* to what Roubaud calls " progressive provisioning," a method which is seen in certain
species of Sphex and Lyroda (*S. politus* and *L. subita,* according to Adlerz) and probably also in *Aphilanthops
frigidus,* according to my observations. But the best examples may be observed among the digger-wasps of the
family Bembicidæ, on which we possess a number of valuable studies by Wesenberg-Lund, Bouvier, Marchal, the
Peckhams, Hartman, Riley, Melander, Ferton, Parker,
Adlerz, the Raus, etc. While our large cicada-killer (*Sphecius speciosus*) provisions its burrows with a single cicada
lays an egg on it and closes the cell, thus practising typical
mass provisioning, some other Stizinæ and many species of
Bembix and allied genera proceed in a different manner.
These insects live in open, sandy places, often in rather
populous and compact congregations, though each female
makes and provisions her own burrow. The prey of each
species of Bembix consists of the common two-winged flies
of her environment, without regard to the species. They
are stung to death but not mutilated. After the burrow
is excavated the wasp kills a small fly and after dislocating
one of its wings, places it on its back on the floor of her cell

and attaches her egg to its sternum. The dislocation of
the wing is supposed by Ferton to be a device for preventing the fly from being turned over by the very delicate
young larva and thus insuring it against injurious contact
with the rough, sandy floor of the cell. The mother collects
flies and brings them into the burrow from day to day, actually increasing the size or number of the victims as they
are needed by the growing and increasingly voracious larva.
At least one European species of Bembix (*B. mediterraneus*), according to Ferton, lays her eggs on the floor of
the cell before bringing in any flies. Instead of flies, the
species of Bicyrtes and Stizus provision their young progressively with bugs or leaf-hoppers, and one of our species
(*Microbembex monodonta*), according to Hartman, Parker
and the Raus, feeds its young on all sorts of small, dead
and dried insects (grasshoppers, beetles, flies, mayflies, ants,
etc.) picked up from the soil.[7]

Among several of the solitary Vespids we find very
similar conditions and these are of more immediate interest
to us because this group of insects has evidently given rise
to the true social wasps. The numerous species of Eumenes and Odynerus (Fig. 24), as well as the allied genera,
either excavate their burrows in the ground, or take possession of the tubular cavities of twigs or the interstices of
walls, or construct exquisite mud cells above ground on the
surfaces of rocks, trees or walls. After the cell is completed
the egg is hung by a filament from its ceiling. Numerous
small, smooth, paralyzed caterpillars are then brought in
and the cell is closed. This is, of course, typical mass provisioning. But Roubaud has observed some very significant modifications of the process in certain Congolese
species. Of one Odynerus (*O. tropicalis*) he gives the following account: " This little Odynerus does not provision

its cells with prey amassed in advance, but nourishes its larvæ from day to day, with small, entire, paralyzed caterpillars, which are always given to the larva in very small numbers, till its growth is completed. The egg is never walled up in the cell with provisions hastily amassed. The

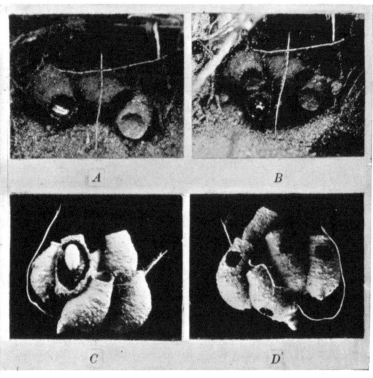

Fig. 24. — Four stages in the mud nest of *Odynerus dorsalis*. *A,* Showing one cell open and being stored with small caterpillars; *B,* Nest on the following day, showing wasp resting in a new cell made on the previous afternoon; *C,* Nest with one cell opened to show the wasp larva feeding on caterpillars; *D,* Same nest, showing holes made by the escaping wasps. (After Carl Hartman.)

wasp, to judge from what I have been able to observe of her educative procedure, after having laid her egg watches it within the cell after the manner of the higher species of Synagris till it hatches. As a rule, prey is brought to the egg only at the moment of hatching or a little before, and usually a single caterpillar, rarely two and never three, is found placed at the disposal of the just-hatched larva. *Pari passu* with the growth of the latter the prey is renewed, but always in small numbers. Sometimes the larva may be seen fasting in the cell while the mother wasp is away in search of prey. Finally, it is only after its feeding has been completed that the larva is immured in the cell. In no case did I observe in closed cells containing larvæ the slightest trace of provisions." Even more interesting are the species of Synagri referred to in this quotation. In several of them Roubaud found the following conditions, representing transitions from mass to progressive provisioning:

The female of *Synagris spiniventris* (cited as *callida* by Roubaud) and *callida* (cited as *sicheliana*) under normal conditions, *i.e.*, when food is abundant, lays an egg in her mud cell, fills it in the course of a few days with small paralyzed caterpillars, sometimes to the number of 60 (!) and then closes it, thus adopting the usual or "banal" method of mass provisioning. When, however, owing to seasonal or climatic conditions, caterpillars are scarce, the female, after ovipositing and guarding the egg for some time, collects a meager provision of small caterpillars for the hatching larva and while it is growing, continues to feed it in the same manner. After the larva has attained three fourths of its adult size, the wasp immures it in its cell with the last supply of provisions. As Bequaert remarks, " in *S. spiniventris*, progressive provisioning is still optional, and one observes all the transitional stages between this behavior and the normal provisioning in mass.

FIG. 25. — Mud nests of *Synagris cornuta* on the thatching of a native hut in the Congo. "Some of these nests show very distinctly the short neck with its slightly widened opening curved to the side and downwards. Such a chimney is built at the entrance of the cells containing eggs or larvæ still nursed by a female." (After J. Bequaert from a photograph by H. O. Lang.)

The mother wasp shows great skill in adapting her habits to the external conditions." According to Roubaud another species of Synagris (*S. cornuta*), proceeds a step further (Figs. 25 and 26). The female, after completing her earthen cell, lays an egg on its floor and when the larva hatches feeds it from day to day with pellets made of a

Fig. 26. — Mud nest of *Synagris cornuta* var. *flavofasciata* with mother wasp. (After J. Bequaert from a photograph by H. O. Lang.)

paste of ground up caterpillars. This is precisely the method employed by the social wasps in feeding their larvæ!

A step in the direction of the social wasps seems also to have been taken by a small group of solitary Vespoids, allied to the Eumeninæ, namely the Zethinæ. These insects, which have been studied recently in Brazil by Ducke, in British Guiana by Howes and in the Philippines by F. X. Williams, have abandoned the use of earth as nest material and employ instead small bits of leaves or moss (Fig. 27). With such vegetable material *Zethus* constructs a beautiful nest with one or several tubular cells, and therefore approaches the social wasps which make their nests of paper,

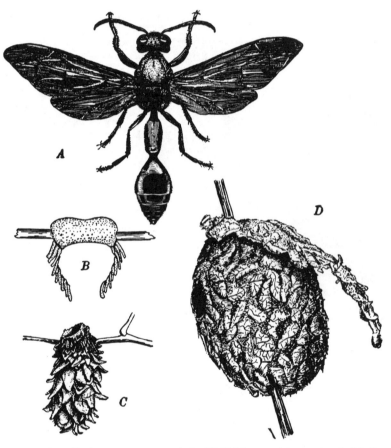

Fig. 27. — *Zethus cyanopterus* of the Philippines and its nest. *A*, Adult wasp x 2; *B*, the beginning of a cell. It is attached to a twig by a mass of well-masticated leaf-bits and the wall of the cell is made of shingled leaf-bits. (Somewhat enlarged); *C*, the first cell of the nest completed x ¾; *D*, a four-cell nest showing roof-like structures and one emergence hole x ¾. (After F. X. Williams.)

a substance consisting of fine particles of wood agglutinated with an oral secretion. The egg is laid loosely in the bottom of the cell and, according to Williams' account of the Philippine *Zethus cyanopterus,* the larva is fed from day to day on small caterpillars, which have been in part eaten by the mother. She faithfully guards the larva and, while it is small and there is still ample room, sleeps in the cell. She closes the latter as soon as the larva is full grown and proceeds to build another.

Each of these cases of progressive provisioning may be regarded as a very primitive family, or society, reduced to its simplest terms, *i.e.,* to a mother and her single offspring. The seasonal or local conditions of the environment, in so far as they affect the abundance or scarcity of prey, have led on the one hand to mass provisioning and therefore to an exclusion of the mother from contact with her growing offspring, and on the other to an establishment of that very contact. This, again, has developed an immediate interest of the mother in her young comparable with what we observe in many birds. Probably this interest is aroused and sustained in the mother wasp by simple, pleasurable, chemical (odor) or tactile stimuli emanating from the egg and larva, but whatever be the nature of the stimuli involved, I believe that we shall have to admit that the egg and the larva have acquired a " meaning " for the mother wasp, and so far as the egg is concerned, this seems to be true even in the species that practice mass provisioning. We noticed that many solitary Vespoids (Eumenes, Odynerus), before they bring in their prey, carefully attach the egg by a string to the ceiling of the cell. This singular performance has been variously interpreted. Fabre and others regard it as a device for preventing the delicate egg from being crushed by the closely packed and sometimes reviving prey, on the same principle that in a crowded room an electric light bulb attached by a cord to the ceiling would be less easily crushed than one rigidly fixed to the walls or the floor. Others regard the filament as a device for keeping the egg free from the occasionally very damp walls of the cell. Ferton has recently shown that *Bembix mediterraneus* glues its slender egg to the floor of the cell in an erect position and with the base carefully supported by a cluster of large sand-grains (Fig. 28 *A*), and that

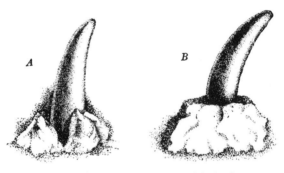

Fig. 28.— Egg of *Bembix mediterraneus* with its base supported by a cluster of large sand-grains. *B*. Egg of *Stizus errans* glued to the upper surface of a carefully selected pebble. (After C. Ferton.)

Stizus errans glues its egg in a similar position to the top of a small, carefully selected pebble placed on the floor of the cell (Fig. 28 *B*). Parker's description of the egg of our *Microbembex monodonta* seems to indicate a condition similar to that of *B. mediterraneus*. In all these cases we seem to have an arrangement for keeping the very easily injured egg as free as possible from contact with the rough, sandy walls of the burrow. But even the Sphecoids and Psammocharids, which practise mass provisioning, attach the egg to a particular part of the victim and in such a position that the hatching larva can attack it at its most vulnerable point.   Ferton, especially, has made a very interesting study of this type of behavior.

The following facts also indicate very clearly that the mother wasp may be aware, not only of sexual differences among her own eggs, but also of the differences in the amount of food required by the resulting larvæ.   Bordage, while investigating the Sphecoids of the Island of Réunion, found that three species, *Pison argentatum, Trypoxylon scutifrons* and *T. errans,* could be readily induced to make their cells in glass tubes placed between the pamphlets of his library.   The Pison, under natural conditions, builds elliptical clay cells and provisions them with spiders, whereas the species of Trypoxylon nest in hollow twigs and the interstices of wall but use the same kind of prey.   All these species adapted themselves to the glass tubes in the same manner.   Each of them plugged the end of the tube with clay and divided the lumen into successive cells by building simple clay partitions across it.   After the cells had been provisioned, Bordage observed that the first of them were longer by half a centimeter and contained more prey than those provisioned later, and he was able to show that the larvæ in the larger, more abundantly provisioned cells produced female, the others male wasps.   Similar observations have also been published by Roubaud on the Congolese *Odynerus (Rhynchium) anceps,* which makes clusters of straight, tubular galleries in clay walls and divides each gallery into several cells by means of clay partitions. In this case also the first cells are much longer than the later, though there is no difference in the quantity of small caterpillars allotted to the different eggs.   But Roubaud was able to prove experimentally that even when the amount of food is so greatly decreased that the larvæ produce adult wasps of only half the normal size, their sex is nevertheless in no wise affected. It would seem therefore that the mother wasp must discriminate between the deposition of a fertilized, female-producing and that of an unfertilized, male-producing egg, and regulate the size of the cell and in some instances also the amount of provisions accordingly.

In the accompanying diagram (Fig. 29), taken from Ducke but somewhat modified, I have indicated the hypothetical family tree of the solitary and social Vespoids. The genera below the heavy horizontal line are solitary, and among them Eumenes and Odynerus seem to be nearest to the original ancestors, because they are very similar to the social forms in having longitudinally folded wings and in other morphological characters. It will be seen that there are six independent lines of descent to the social forms above the heavy line and that the genera plotted at different levels represent various stages of specialization as indicated by the nature of the materials and types of structure of the nests. With the doubtful exception of a few Stenogastrinæ, all the social wasps make paper nests consisting wholly or in part of one or more combs or regular hexagonal cells, in which a number of young are reared simultaneously.

Authorities on the classification of the social wasps now divide them into five subfamilies, namely the Stenogastrinæ, which are confined to the Indomalayan and Australian Regions, the Ropalidiinæ, confined to the tropics of the Old World, the Polistinæ, which are cosmopolitan, the Epiponinæ, possibly comprising two independent lines of descent from Eumenes-like and Odynerus-like ancestors respectively and constituting a large group, mostly confined to tropical America, with a few species in the Ethiopian, Indomalayan, Australian and North American regions,

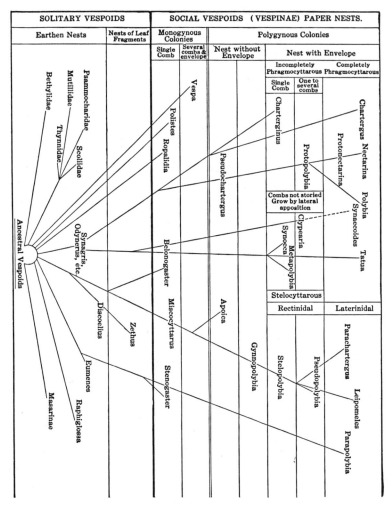

Fig. 29.—Phylogenetic tree of the various genera and families of Vespoids. (After Ducke, with modifications.)

and the Vespinæ, which are recorded from all the continents except South America and the greater portion of Africa south of the Sahara. These five families may be briefly characterized before considering some of the peculiarities of social organization common to most or all of them.

1. The Stenogastrinæ evidently represent a group of great interest, because they form a transition from the solitary to the social wasps, but unfortunately our knowledge of their habits is very incomplete. F. X. Williams has recently published observations on four Philippine species, and though his account is fragmentary, it nevertheless reveals some peculiar conditions. He shows that the single genus of the subfamily, Stenogaster, includes both solitary and social forms and that all of them exhibit a mixture of primitive and specialized traits. The species all live in dark, shady forests and make very delicate, fragile nests with particles of decayed wood or earth. *S. depressigaster* (Figs. 30 *E* and *F*) hangs its long, slender, cylindrical nests to a pendent hair-like fungus or fern. The structure consists of tubular, intertwined galleries and cells, with their openings directed downwards. The colony comprises only a few individuals, probably the mother wasp and her recently emerged daughters. The eggs are attached to the bottoms of the cells as in all social wasps and the larvæ are fed from day to day with a gelatinous paste, which Williams believes may be of vegetable origin. In the cells the older larvæ and the pupæ hang head downwards. Another social species, *S. varipictus,* constructs a very different nest, consisting of cells made of sandy mud mixed perhaps with particles of decayed wood and attached side by side in groups to the surfaces of rocks and tree trunks (Fig. 30 *G*). In this case also the cell-openings are directed downward. A nest may consist of thirty or more cells in several rows. There are only a few wasps in a colony, and when the larvæ are full-grown the cells are sealed up by the mother as in the solitary wasps. But after the young have emerged the cells may be used again as in many of the social species.

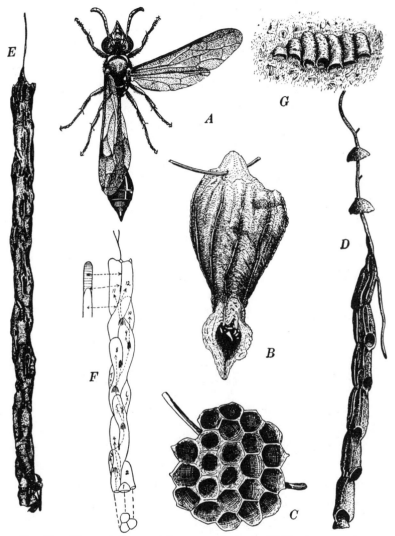

Fig. 30.—Nests of Stenogastrine wasps from the Philippines. *A*, *Stenogaster micans* var. *luzonensis*, female; *B*, Completed nest of same; *C*, Nest with only the basal portion completed; *D*, Nest of *Stenogaster* sp., with umbrella-like "guards"; *E*, Nest of *S. depressigaster*; *F*. diagram of same showing arrangement of cells and passage-ways. The numbers indicate the cells. The tops of the passage-ways are shown in two planes by series of parallel lines; *G*, Nest of *S. varipictus* on the bark of a tree. (After F. X. Williams.)

Williams describes and figures the nests of two solitary species, one an undetermined form, the other identified as *S. micans* var. *luzonensis*. The nest of the former (Fig. 30 *D*) is suspended, like that of *depressigaster*, from some thin vegetable fiber and appears to consist of particles of decayed wood. It is a beautiful, elongate structure of seven tubular, ribbed cells arranged in zigzag series with their openings below and two peculiar umbrella-like discs around the supporting fiber. These discs " remind one a good deal of the metal plates fastened to the mooring lines of vessels and serving as rat guards. Their function in the case of the nest may be an imperfect protection from the ants, or perhaps they may serve as umbrellas, though neither they nor the cells are strictly rain proof." They may possibly be rudiments of the nest envelopes which are so elaborately developed in many of the higher social wasps. The mother wasp attends to several young simultaneously, and when their development is completed seals up the cells. *S. micans* var. *luzonensis* (Fig. 30 *B*) makes the most remarkable nest of all. It is attached to some pendent plant filament under an overhanging bank or under masses of dead leaves supported by twigs or vines and is made of " moist and well-decayed wood chewed up into a pulp and formed into delicate paper which is not rain proof." The basal portion of the nest (Fig. 30 *C*) is a single comb of about 20 regular, hexagonal cells, enclosed in a pear-shaped covering which is longitudinally grooved and ribbed on the outside and constricted below to form a filigree-work, funnel-like aperture surrounded on one side by a spear-shaped expansion. This species seems also to have been observed in Ceylon by E. E. Green, who remarks that " the nest seems to be the property of one pair only " of wasps. Two other species, *S. nigrifrons* of Burma and *melleyi* of Java, are also recorded as social. They make nests consisting of a few pendent, hexagonal-celled combs attached to one another by slender pedicels. All of the descriptions indicate that the colonies of the social species of Stenogaster

Fig. 31. — Suspended and naked comb of a very primitive African Epiponine wasp, *Belonogaster junceus*, with young cells above and old cells containing larvæ below; natural size. Most of the wasps have been removed but two are seen bringing food-pellets to the larvæ. (Photograph by E. Roubaud.)

must consist of a very few individuals, and there is nothing to show that the female offspring differ in any way from their mother or that they assist in caring for the brood. Even in the case of *S. varipictus,* Williams remarks: " In a small way, it seems to be a social wasp; one to several insects attend to a cell group. It may be, however, that each female has her own lot of cells in this cell group." Future investigations may show that none of the species of Stenogaster is really social in the same sense as are the four other subfamilies, though they approach the definitely social forms in using paper in the construction of the nest, in sometimes making combs of regular hexagonal cells and in caring for a number of larvæ at the same time.

2. The Epiponinæ are a large and heterogeneous group, comprising a much greater number of genera (23) than any other subfamily of social wasps, and ranging all the way from very primitive forms like Belonogaster to highly specialized forms like Chartergus and Nectarina. Great differences are also apparent in the architecture of the nest, which in the more primitive genera consists of a single naked comb of hexagonal cells attached to some support by a peduncle (Fig. 31), and in the more advanced forms of a single comb or of several combs superimposed on one another and enclosed in an envelope with an opening for ingress and egress (Figs. 32 and 33). The combs are in some cases pedunculate (stelocyttarous), in others attached directly to the support or to the envelope (phragmocyttarous). In nearly all cases the nest is made entirely of paper, but in a few tropical American species some clay may be added. It is always above ground and attached to the branches or leaves of trees, to the underside of some shelter (roofs, banks, etc.). In primitive forms like Belonogaster (Fig. 31), as a rule, a single fecundated female starts the nest by building a single pedunculate cell and then gradually adding others in circles concentrically to its periphery as the comb grows, but not infrequently the foundress may be joined by other females before the work

has progressed very far. Each larva is fed with pellets of malaxated caterpillars till it is full grown, when it spins a convex cap over the orifice of its cell and pupates. The emerging females are all like the mother in possessing well-developed ovaries and in being capable of fecundation. In other words, all the females of the colony are physiolog-

FIG. 32. — Nest of *Polybioides tabida* from the Congo, with the involucre partly removed. (After J. Bequaert from a photograph by H. O. Lang.)

ically equal, and even such differences in stature as they may exhibit have no relation to fertility. The colonies are small, the nests having usually only about 50 to 60 cells, rarely as many as 200 to 300. In larger colonies there is a certain rude division of labor since the older females devote themselves to egg-laying, the younger to foraging for food

Fig. 33. — A. Nest of *Polybioides melæna* of the Congo. B. The same partly destroyed, showing the pendent combs, which have cells on both sides. (After J. Bequaert, from a photograph by H. O. Lang).

and nest materials and the recently emerged individuals to feeding the larvæ and caring for the nest. The males, too, remain on the comb, but behave like parasites and exact food whenever it is brought in by foraging females. Belonogaster is described as a polygynous wasp because each of its colonies contains a number of fecundated females. When it has reached its full development the females leave in small companies and found new nests either singly or together. This phenomena is known as "swarming" and occurs only in the wasps of the tropics where it seems to be an adaptation to the favorable climatic conditions. In the higher South American genera of Epiponinæ, however, the females are not all alike but are differentiated into true females, or queens, *i.e.*, individuals with well-developed ovaries and capable of fecundation, and workers, *i.e.*, females with imperfectly developed ovaries and therefore sterile or capable only of laying unfertilized, male-producing eggs. Many of these wasps, according to H. and R. von Ihering and Ducke, are polygynous and regularly form new colonies and nests by sending off swarms of workers with one or two dozen queens. The colonies often become extremely populous and comprise hundreds or even thousands of individuals. Some of the species (Nectarina, Polybia) have a habit of storing a considerable amount of honey in their combs, while others are known to capture, kill and store within the nest envelope, and even in the combs, quantities of male and female termites or male ants as a supply of food to be drawn on when needed.[8]

3. The Ropalidiinæ are a small group of only three genera, the best known of which is Ropalidia. These are primitive wasps which build a single naked comb like that of Belonogaster and feed their young with pellets of malax-

ated insects. The colonies are small and polygynous, but according to Roubaud, true workers can be distinguished, though they are few in number compared with the true females. Swarming seems to occur in some species.

4. The Polistinæ are represented by only two genera. One of these, Polistes, is cosmopolitan and, like Ropalidia and Belonogaster, makes a single, naked comb, suspended by a central or eccentric peduncle to the underside of some shelter. As there are several common species in Europe and the United States, the habits of the genus are well-known. The nest is usually established and in its incipient stages constructed by a single female, or queen. A certain number of her offspring are workers though they seem often to lay male-producing eggs. True females are rather numerous in the colonies of some species, which may therefore be regarded as polygynous, and some of the tropical forms may, perhaps, swarm. In temperate regions, however, the Polistes colony is an annual development and usually not very populous. The young females are fecundated in the late summer and pass the winter hidden away under bark or in the crevices of walls, whence they emerge in the spring to found new colonies. Several of the species, even in temperate regions, are known to store small quantities of honey in their combs.

5. Like the Polistinæ, the subfamily Vespinæ includes only two genera, Vespa and Provespa. The species of the former, the only genus besides Polistes that occurs in the north temperate zone, are the largest and most typical of social wasps. So far as known the species are strictly monogynous. The nest, founded by a single female, consists at first of a small pendent comb, like that of Polistes, but while there are still only a few cells a more or less spherical envelope is built around it. The eggs first laid

produce workers, which are much smaller than the mother and incapable of fecundation. They remain with the parent, enlarge the comb and envelope and, to accomodate the rapidly increasing brood, build additional combs in a series from above downward, each new comb being supported by one or more peduncles attached to the comb above it (stelocyttarous). At first large numbers of workers are produced, but later in the summer males and females appear. Owing to the greater size of the females, the cells in which they are reared are considerably larger than the worker cells. After the mating of the males and females the colony perishes, with the exception of the fecundated females, which hibernate like the females of Polistes and during the following spring found new colonies. In the Vespinæ, therefore, a very distinct worker caste has been developed, though its members occasionally and perhaps regularly lay male-producing eggs. The species of Vespa are usually divided into two groups, one with long, the other with very short cheeks. In Europe and North America the long-cheeked forms as a rule build aerial nests above ground, the short-cheeked forms in cavities which they excavate in the ground. The colonies may often be very populous by the end of the summer (3,000 to 5,000 individuals).

After this hasty sketch of the five subfamilies of the social wasps we may consider a few of their fundamental behavioristic peculiarities, especially the trophic relations between the adults and larvæ, the origin of the worker caste, its ultimate fate in certain parasitic species and the question of monogyny and polygyny. In all these phenomena we are concerned with effects of the food-supply and therefore of the external environment.

The feeding of the larvæ by Vespa and Polistes queens

and workers with pellets made of malaxated portions of caterpillars, flies or other insects has often been described and can be readily witnessed in any colony kept in the laboratory. The hungry larvæ protrude their heads with open mouths from the orifices of the cells, like so many nestling birds, and when very hungry may actually scratch on the walls of the cells to attract the attention of the workers or their nurses. The feeding is not, however, a one-sided affair, since closer observation shows that the wasp larva emits from its mouth drops of sweet saliva which are eagerly imbibed by the nurses. This behavior of the larvæ has been observed in all four subfamilies of the higher wasps by du Buysson, Janet and Roubaud. Du Buysson says that the larvæ of Vespa " secrete from the mouth an abundant liquid. When they are touched the liquid is seen to trickle out. The queen, the workers and the males are very eager for the secretion. They know how to excite the offspring in such a way as to make them furnish the beverage." And Janet was able to prove that the secretion is a product of the salivary, or spinning glands and that it flows from an opening at the base of the lower lip. " This product," he says, " is often imbibed by the imagines, especially by the just emerged workers and by the males, which in order to obtain it, gently bite the head of the larva." Most attention has been bestowed on this reciprocal feeding by Roubaud, from whose interesting account of Belonogaster, Ropalidia and Polistes I take the following paragraphs:

" All the larvæ from birth secrete from a projection of the hypopharynx, on the interior surface of the buccal funnel, an abundant salivary liquid, which at the slightest touch spreads over the mouth in a drop. All the adult wasps, males as well as females, are extremely eager for this salivary secretion, the taste of which is slightly sugary. It is easy to observe, especially in Belonogaster, the insistent demand for this larval product and the tactics employed to provoke its secretion.

" As soon as a nurse wasp has distributed her food pellet among the various larvæ, she advances with rapidly vibrating wings to the opening of each cell containing a larva in order to imbibe the salivary drop that flows abundantly from its mouth. The method employed to elicit the secretion is very easily observed. The wing vibrations of the nurse serve as a signal to the larva, which, in order to receive the food, protrudes its head from the orifice of the cell. This simple movement is often accompanied by an immediate flow of saliva. But if the secretion does not appear the wasp seizes the larva's head in her mandibles, draws it toward her and then suddenly jams it back into the cell, into which she then thrusts her head. These movements, involving as they do a stimulation of the borders of the mouth of the larva, compel it to secrete its salivary liquid.

" One may see the females pass back and forth three or four times in front of a lot of larvæ to which they have given nutriment, in order to imbibe the secretion. The insistence with which they perform this operation is such that there is a flagrant disproportion between the quantity of nourishment distributed among the larvæ by the females and that of the salivary liquid which they receive in return. There is therefore actual exploitation of the larvæ by the nurses.

" The salivary secretion may even be demanded from the larva without a compensatory gift of nourishment, both by the females that have just emerged and by the males during their sojourn in the nest. The latter employ the same tactics as the females in compelling the larvæ to yield their secretion. They demand it especially after they have malaxated an alimentary pellet for themselves, so that there is then no reciprocal exchange of nutritive material.

" It is easy to provoke the secretion of the larvæ artificially. Merely touching the borders of the mouth will bring it about. The forward movement of the larvæ at the cell entrance, causing them to protrude their mouths to receive the food pellet, is also easily induced by vibrations of the air in the neighborhood of the nest. It is only necessary to whistle loudly or emit shrill sounds near a nest of Belonogaster to see all the larvæ protrude their heads to the orifice of the cells. Now it is precisely the vibrations of the air created by the rapid agitation of the bodies of the wasps and repeated beating of their wings that call forth these movements, either at the moment when food is brought or for the purpose of obtaining the buccal secretion which is so eagerly solicited."

Roubaud has called the interchange of food here described " œcotrophobiosis," but for reasons which I cannot stop to discuss, I prefer to use the word " trophallaxis." It will be seen that the larvæ have acquired a very definite meaning for the adult wasps of all castes and that through trophallaxis very close physiological bonds have been established, which serve to unite all the members of the colony, just as the nutritive blood stream in our bodies binds all the component cells and tissues together. We found that even in forms like *Synagris cornuta* the larva has acquired a meaning for the mother. In this case Roubaud has shown that the mother while malaxating the food-pellet herself imbibes its juices before feeding it to the larva, and that " the internal liquids having partly disappeared during the process of malaxation, the prey is no longer, as it was in the beginning, soft and juicy and full of nutriment for the larva. It is possible, in fact, to observe that the caterpillar *pâté* provided by the *Synagris cornuta* is a coarse paste which has partly lost its liquid constituents. There is no exaggeration in stating that such food would induce in larvæ thus nourished an increase of the salivary secretion in order to compensate for the absence of the liquid in the prey and facilitate its digestion." It is here that the further development to the condition seen in Belonogaster and other social wasps sets in. The mother finds the saliva of the larva agreeable and a trophallactic relationship is established. As Roubaud says, "the nursing instinct having evolved in the manner here described in the Eumenids, the wasps acquire contact with the buccal secretion of the larva, become acquainted with it and seek to provoke it. Thence naturally follows a tendency to increase the number of larvæ to be reared simultaneously in order at the same time to satisfy the urgency of oviposition and to profit by the greater abundance of the secretion of the larvæ."

As I shall endeavor to show in my account of the ants and termites, trophallaxis is of very general significance in the social life of insects. It seems also to have an important bearing on the development of the worker caste. Both queens and workers arise from fertilized eggs, and the differences between them are commonly attributed to the different amounts of food they are given as larvæ. There seems to be much to support this view in the social wasps. As Roubaud points out in the passages quoted, the larvæ are actually exploited by the adult wasps to the extent of being compelled to furnish them with considerable quantities of salivary secretion, often out of all proportion to the amount of solid food which they receive in return. Owing to this expenditure of substance and the number of larvæ which are reared simultaneously, especially during the earlier stages of colony formation, they are inadequately nourished and have to pupate as rather small individuals, with poorly developed ovaries. Such individuals therefore become workers. This inhibition of ovarial development, which has been called "alimentary castration," is maintained during the adult life of most workers by the exigencies of the nursing instincts. The workers have to complete and care for the nest, forage for food and distribute most of it among their larval sisters. All this exhausting labor on slender rations tends to keep them sterile. In other words, "nutricial castration" (derived from *nutrix,* a nurse) to use Marchal's terms, takes the place in the adult worker of the alimentary castration to which it was subjected during its larval period. It is only later in the development of the colony, when the number of workers and consequently also the amount of food brought in have considerably increased, and the labor of foraging and nest construction have correspondingly decreased for the individual worker, that the larvæ can be more copiously fed and develop as fertile females, or queens. At that season, too, some of the workers may develop their ovaries, but as the members of the worker caste are incapable of fecundation, they can lay only male-producing eggs. That this is not the whole explanation of the worker caste will appear when we come to consider the much more extreme conditions in the ants and termites, but it may suffice to explain the conditions in the social wasps and social bees.

Parasitism is another phenomenon which seems to indicate that a meager or insufficient diet is responsible for the development of the worker caste. Although parasitic species are much more numerous among the bees and ants, I will stop to consider very briefly a few of those known to occur among the wasps. A parasite is, of course, an organism that is able to secure abundant nourishment for itself or its offspring by appropriating the food-supply that has been laboriously stored or assimilated by some other organism. The various parasitic solitary wasps, such as the species of Ceropales, among the spider-storing Psammocharidæ, all substitute their own young for the young of their hosts in order that the larvæ may come into undisputed possession of the stored provisions. Among the social wasps there are only two parasitic species, *Vespa austriaca* and *V. arctica*. The former has long been known in Europe where it lives in the nests of *V. rufa*. Recently Bequaert and Sladen have found *austriaca* in the United States, British America and Alaska, but its Cisatlantic host is still unknown, though believed to be *V. consobrina*. *V. arctica*, as Fletcher, Taylor and I have demonstrated, lives in the nests of our common yellow jacket (*V. diabolica*). Now both *austriaca* and *arctica* have completely lost the worker caste so that they are represented only by males and fertile females. They were at one time undoubtedly non-parasitic like their present hosts, but are now reared and fed by the workers of the latter like their own more favored sexual forms. As a result of such nurture what were once independent social insects with two female forms have actually reverted to the status of solitary forms with only one type of female.

In conclusion the conditions of monogyny and polygyny in the higher social wasps may be briefly considered. It was shown that the Vespinæ and at least most of the Polistinæ are monogynous, their colonies being annual developments begun by a single fecundated queen, and that they perish at the end of the season, with the exception of the annual brood of queens, which after fecundation hibernate and start new colonies during the following spring. Many of the tropical Epiponinæ and Ropalidiinæ, however, are polygynous and the former often form large perennial colonies which from time to time send off swarms consisting of numerous fecundated females or of such females accompanied by workers to found new colonies. This behavior is evidently as perfect an adaptation to the continuously favorable food and temperature condition of the tropics as is that of Vespa and Polistes to the pronounced seasonal vicissitudes of the temperate regions. There has been a difference of opinion among the authorities as to whether monogyny or polygyny represents the more primitive phylogenetic stage among the social wasps. The great majority of these insects are tropical, and probably even Vespa and Polistes were originally inhabitants of warm regions and invaded temperate Eurasia and North America during postglacial times. The monogyny still exhibited by these wasps in the tropics may have been acquired there as an adaptation to the wet and dry seasons, and this adaptation may have enabled them the more easily to adjust themselves to the warm and cold seasons of more northern regions. H. and R. von Ihering and Roubaud may therefore be right in maintaining that polygyny is the more primitive condition. Their view is also supported by the fact that in the polygynous genera the worker caste is either still absent (Belonogaster) or very feebly developed and constitutes only a small percentage of the female personnel of the colony. We might, perhaps, say that our species of Vespa and Polistes each year produce a swarm of females and workers but that the advent of cold weather destroys the less resistant workers and permits only the dispersed queens to survive and hibernate till the following season.

We shall find precisely the same differences between monogyny and polygyny in the social bees of temperate and tropical regions, and somewhat analogous conditions among the ants, although their polygyny may be secondarily derived from monogyny. It would seem that swarming must be a phenomenon which occurs as a rule when the environment is unfavorable or the colony has grown to such dimensions as to outrun its food-supply so that emigration of portions of its population becomes imperative.

LECTURE III

BEES SOLITARY AND SOCIAL

PART 1

To those who are not entomologists the word " bee " naturally signifies the honey-bee, because of all insects it has had
the most delightful, if not the longest and most intimate
association with our species. Of course, the key to the understanding of this association is man's natural appetite
or craving for sweets, and the fact that till very recently
honey was the only accessible substance containing sugar
in a concentrated form. It is not surprising, therefore, that
man's interest in the honey-bee goes back to pre-historic
times. He was probably for thousands of years, like the
bears, a systematic robber of wild bees till, possibly during
the neolithic age, he became an apiarist by enticing the
bees to live near his dwelling in sections of hollow logs,
empty baskets or earthen vessels. Savage tribes keep bees
to-day, and within their geographic range we know of no
people who have not kept them. They figure on the Egyptian monuments as far back as 3500 B.C., and we even
know the price of strained honey under some of the Pharaohs. It was very cheap — only about five cents a quart.

The keeping of the honey-bee could not fail to excite the
wonder and admiration of primitive peoples. It was at
once recognized as a privileged creature, for it lived in
societies like those of man, but more harmonious. Its sustained flight, its powerful sting, its intimacy with flowers

and avoidance of all unwholesome things, the attachment
of the workers to the queen — regarded throughout antiquity as a king — its singular swarming habits and its
astonishing industry in collecting and storing honey and
skill in making wax, two unique substances of great value
to man, but of mysterious origin, made it a divine being,
a prime favorite of the gods, that had somehow survived
from the golden age or had voluntarily escaped from the
garden of Eden with poor fallen man for the purpose of
sweetening his bitter lot.   No wonder that the honey-bee
came in the course of time to symbolize all the virtues —
the perfect monarch and the perfect subject, together constituting the perfect state through the exercise of courage,
self-sacrifice, affection, industry, thrift, contentment, pur-
ity, chastity — every virtue, in fact, except hospitality,
and, of course, among ancient peoples bent on maintaining
their tribal or national integrity, the fact that bees will not
tolerate the society of those from another hive was interpreted as a virtue.

With the passing centuries the bee became the center of
innumerable myths and superstitions.   It was supposed to
have played a rôle in the lives of all the more important
Egyptian, Greek and Roman divinities.   Among the Latins
it even had a divinity of its own, the goddess Mellonia.
Medieval Christians seem to have been quite as eager to
show their appreciation of the insect.   While the housefly
had to be satisfied with the patronage of Beelzebub and the
ant was given so obscure a patron saint as St. Saturninus,
the honey-bee enjoyed the special favor of the Virgin or
was even made the " *ancilla domini*," the maid-servant of
the Lord.   Those who represented the divinity on earth, of
course, added the honey-bee to their insignia.   It appears
on the crown of the Pharaohs as the symbol of Lower

Egypt, on the arms of popes and on the imperial robes of the Napoleons. Among the ancients the behavior of bees was supposed to be prophetic and the insect thus naturally became associated with Apollo, the Delphic priestess, the Muses and their protegés, the poets and orators. Honey and wax were early believed to have medicinal and magical properties and were, of course, used for sacrificial purposes. Their ritual value is apparent also in the Christian cult, for honey was formerly given to babies during baptism and the tapers of our churches are supposed to be made of pure bees' wax (" *nulla lumina nisi cerea adhibeantur* ").[1]

Among the many myths that have grown up around the honey-bee, that of the " bugonia " may be considered more fully, because it shows how entomology may throw light on questions that have puzzled and distracted the learned for centuries. For nearly three thousand years people believed that the decomposing carcass of an ox or bull can produce a swarm of bees by spontaneous generation. The myth evidently started in Egypt and appears in a distorted form among the Hebrews, among whom, however, it is a dead lion in which Samson finds the honey-comb. Among the Greeks and Romans it becomes more elaborate, and Virgil, in the fourth book of the Georgics, and many other authors give precise directions for the killing and treatment of the ox if the experiment is to be successful. The medieval writers repeat what they read in the classics or invent more fantastic accounts. It was not till the eighteenth century that Réaumur showed that what had been regarded as bees issuing from the decomposed ox carcass must have been large two-winged flies of the species now known as *Eristalis tenax,* which breed in great numbers in carrion and filth and look very much like worker bees. The history of this myth of the oxen-born bees has been more adequately discussed by a distinguished dipterist, Baron Osten Sacken.[2] He remarks that "the principal factor underlying the whole intellectual phenomenon we are inquiring into is the well-known influence which prevails in all human matters, and this factor is *routine*." "Thinking is difficult, and acting according to reason is irksome," said Goethe. People see and believe in what they see, and the belief easily becomes a tradition. It may be asked: If those people had that belief, why did they not try to verify it by experiment, the more so as an economical interest seemed to be connected with it? The answer is that they probably did try the experiment, and did obtain *something* that looked like a bee; but that there was a second part of the experiment, which, if they ever tried it, never succeeded, and that was, to make that bee-like something produce honey. If they did not care much about this failure, and did not prosecute the experiment any further, it is probably because, in most cases, they found that it was much easier to procure bees in the ordinary way. That such was really the kind of reasoning which prevailed in those times clearly results from the collation of the passages of ancient authors about the "*Bugonia.*"[3]

It would seem that the strange vitality of the bugonia myth during so many centuries must have been due to some keen emotional factor or religious conviction deeper than the mere inertia of routine thinking to which Osten Sacken refers. Let us work backwards from the golden bees embroidered on the state robes of Napoleon I. and supposed to symbolize his official descent from Charlemagne, who is said to have worn them on his coat of arms. It is probable that the fleur-de-lys, which also figure on his arms and those of the later French kings, are really conventionalized bees and not lilies, spear-heads or palm trees with horns or amulets attached, as some archeologists have asserted, and that Charlemagne derived his bees from one of the first kings of the Salian Franks, the father of Clovis, Childeric I., who died A.D. 481. In 1653 the tomb of this monarch was opened at Tournay, in Flanders, and found to contain a number of objects which indicated that he had been initiated into the cult of Mithra, that soldiers' religion which had been so widely diffused by the Romans over Gaul, Britain and Germany during the first centuries of our era and had come so near to supplanting Christianity. Among the objects taken from Childeric's tomb were a golden bull's head and some 300 golden bees, set with precious stones and provided with clasps which held them to the king's mantle. Now the numerous Mithraic monuments that have been unearthed in many parts of the Roman empire show as their central figure Mithra slaying a bull surrounded by several symbolic animals, one of which is the bee. It is known also that honey was used in the initiation rites of Mithra, who was an oriental sun-god like the Hebrew Samson, the Phœnician Melkart and the Greek Hercules. From the blood of the slain bull, a symbol of the inert earth fertilized by the sun's rays, the animal world was supposed to have arisen by spontaneous generation. The bee would seem, therefore, to be one of the symbols of this renewal of life and to recall the epiphanies of many other sun and vegetation gods among the Greeks and Asiatic peoples, such as Adonis, Attis and Dionysus, or Bacchus, who as Dionysus Briseus, the "squeezer of honeycomb," was by some regarded as the god of apiculture. But the bugonia myth can be traced still further back to the Apis cult of the Egyptians. The bull Apis was believed to be an incarnation of the sun-god Osiris and to represent the renewal of life. His son Horus is another sun-god, and it is interesting to note that one of his symbols is the fleur-de-lys, which signifies resurrection. That this is the true meaning of the bugonia myth is indicated also by the magical directions given by Virgil and others for slaying the ox and caring for his carcass. The animal must be carefully chosen and in the spring, when the sun is in the sign of the bull, clubbed to death or suffocated by having the apertures of his body stuffed with rags — obvious precautions to prevent the ox's vitality from escaping so that it may be conserved for the generation of the swarm of bees. The ancients seem to have had an inkling of the parthenogenesis of the honey-bee, since many of them state that, unlike other animals, it never mates. This belief, too, served to connect the bee with the various sun and vegetation gods, all of whom, including the bull Apis, were born of virgins. Thus it will be seen that the bee became the symbol of the ever-recurring resurrection, or renewal of life in general, and hence probably also of the second birth of the initiate into such cults as those of Mithra. Unfortunately there were among the ancients no entomologists to point out to the religious enthusiasts that they had mistaken a common carrion fly for the honey-bee and had therefore chosen a wrong symbol.

I have dwelt on this myth because it is such a good example of the bad observation and worse conjecture that have clouded our knowledge of the honey-bee. Even such pioneer observers as Swammerdam, Réaumur and François Huber in the seventeenth and eighteenth centuries and Dzierzon, Leuckart, von Siebold and von Buttel-Reepen in more recent times have had difficulty in clearing a path through the jungle of superstitions and speculations that have grown up around the insect during the past five thousand years. And to-day many of our scientific treatises contain vestiges of these unbridled fancies. Another obstacle to a clear understanding of the honey-bee is the very abundance of the literature. There must have been libraries devoted to it among the ancients, for even Carthage had her celebrated apiarists. Some notion of the present conditions may be gleaned from Dr. E. F. Phillips' statement that the Bureau of Entomology at Washington has a working bibliography of 20,000 titles on the honey-bee. This does not, of course, include a great number of bellettristic works like Virgil's Georgics, Maeterlinck's " Vie des Abeilles " and Evrard's " Mystère des Abeilles."

Greatest of all the sources of a misunderstanding of the honey-bee is the fact that although it is a very highly specialized and aberrant insect, it has been regarded as a paragon in the light of which the social organizations of all other insects are to be interpreted. Its evolutionary interpretation has therefore encountered the same obstacles as that of man, for the honey-bee bears much the same relation to other bees that man does to the other mammals; and just as man's obstinate anthropocentrism has retarded his understanding of his own history and nature, so the apicentrism of the observers of the honey-bee has tended to distort our knowledge, not only of other social insects but of the honey-bee itself. It is necessary, therefore, to relegate the insect to its proper place at the end of a long series of developments. I shall return to it at the end of the lecture.

As classified by the entomologists, the bees comprise about 10,000 described species and occur in all parts of the world. In Europe alone there are some 2,000 species and our North American forms, when thoroughly known, will probably be found to be even more numerous. Less than 500, or 5 per cent., of the 10,000 species are social and belong to only five genera — Trigona, Melipona, Bombus, Psythirus and Apis — the remainder being solitary forms of many genera, several of which are very large and widely distributed. For more than a century talented entomologists have studied the bees intensively but have been unable to work out any generally acceptable grouping of the various genera. Whether these insects are to be regarded as a superfamily (Apoidea), comprising several families, or as a single family (Apidæ), comprising a number of subfamilies, seems to depend on the individual investigator's more radical or more conservative frame of mind.

The bees, taken as a whole, are properly regarded merely as a group of wasps, which have become strictly vegetarian and feed exclusively on the pollen and nectar of flowers. They are, in a word, merely flower-wasps — " Blumenwespen," as they are called by some German entomologists. A recent authority, Friese, believes that they are descended directly from at least two different ancestral groups of Sphecoid solitary wasps, one of which includes genera like Passalœcus and leads up to Prosopis and other primitive bees, while the other comprises Tachytes-like forms and leads up to the higher bees. It should be noted that a third ancestral group of Vespoids, allied to the Eumenid wasps evidently gave rise to the Masarinæ, which are also flower-wasps and in their habits closely resemble the solitary bees.[4]

Their very long and intimate association with the flowers has left its stamp on all the organs and habits of the bees, and botanists believe that a great many flowers have been modified in structure, arrangement, color and perfume in adaptation to the bees and for the purposes of insuring cross-pollination. Limitations of time prevent me from dwelling on the vast and fascinating subject of these relationships, though they belong to that order of interorganismal cooperation which I have called biocœnobic. Nor can I stop to dwell on our great debt to the bees for the pollination of our fruit trees and other economic plants. Something must be said, however, concerning the anthophilous adaptations of the insects themselves. It is evident that only insects with well-developed wings, with large, finely faceted eyes and well-developed antennæ, furnished with extremely delicate tactile and olfactory sense-organs, could have acquired such intimate relations to the flowers. And since the bees not only collect but transport the pollen and nectar we find some very interesting structures developed for these particular functions. Two pairs of mouth parts, the maxillæ and especially the tongue, are peculiarly modified for lapping or sucking up the nectar. In the more primitive bees that visit flowers with exposed nectaries these parts are short and much like those of the wasps, whereas in more specialized species that visit flowers with nectaries concealed in long tubes the tongue is greatly elongated. In some tropical bees the organ may be even longer than the body (Fig. 34). In order to store the nectar while it is being transported to the nest, the crop, or anterior portion of the alimentary tract, is large, bag-like and distensible and its walls are furnished with muscles which enable the bee to regurgitate its content. This is known as honey, because the nectar, during its sojourn in the crop, is mixed with a minute quantity of a ferment, or enzyme, presumably derived from the salivary glands, and undergoes a chemical change, its sucrose, or cane sugar being converted into invert sugars (levulose and dextrose). Even more striking are the adaptations for collecting and carrying the pollen. The whole surface of the bee's body is covered with dense, erect hairs, which, unlike those of other insects, are branched, plumose, or feather-like and easily hold the pollen grains till the bee can sweep them together by combing itself with its legs (Fig. 35). Many bees thus bring the pollen together into masses moistened with a little honey and attach them to the outer surfaces of the tibiæ and metatarsi of the hind legs (Figs. 37 and 38). These parts are peculiarly broadened and provided with long hairs to form a special pollen-

FIG. 34. — A long-tongued Neotropical bee (*Eulæma mussitans*). About twice natural size. (Original.)

basket or corbicula (Fig. 36). In other bees the pollen is swept to the ventral surface of the abdomen, where there are special hairs for holding it in a compact mass. The bees of the former group are therefore called " podilegous," the latter " gastrilegous." That these various structures, *i.e.,* the general body investment of plumose hairs and the modifications of the hind legs or venter are special adaptations for pollen collection and transportation is proved by certain interesting exceptions. Thus the small bees of the

very primitive genus Prosopis look very much like diminutive wasps; they have naked bodies and appendages and
their hind legs are not modified. But these bees swallow
the pollen as well as the honey and carry both in their
crops. Then there is a long series of genera of parasitic
bees which lay their eggs in the nests of the industrious
species and on this account do not need any collecting or
transporting apparatus. Such bees are more or less naked

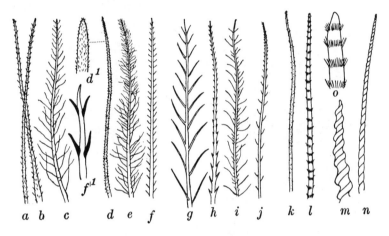

FIG. 35. — Hairs of various bees. *a-f,* of bumble-bees; *g-j,* of *Melissodes* sp.; *k-n,* of *Megachile* sp. (After John B. Smith.)

and their tibiæ have returned to the simple structure seen in
the wasps. And, of course, since male bees in general do
not have to collect pollen we find that they, too, show considerable reduction in the hind legs as compared with the
cospecific females.

There are great differences among the bees in the range
of their attachment to the flowers. Some, like the honey-
bee and the bumble-bees, visit all sorts of flowers and are
therefore called polytropic, whereas others, the so-called
oligotrophic species, may confine their attentions to the

flowers of a very few plants or even to those of a single species. The oligotropic are probably derived from polytropic bees which have found it advantageous to avoid competition with other species and to make their breeding

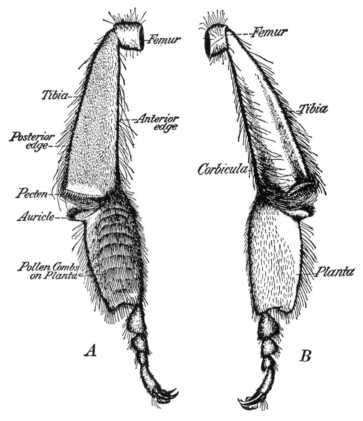

Fig. 36. — *A*, Inner surface of the left hind leg of a worker honeybee; *B*, Outer surface of the same. (After D. B. Casteel.)

season coincide with the blooming period of a single plant. A good example is one of our small black bees, *Halictoides novæ-angliæ* which at least in New England visits only the purple flowers of the pickerel weed, *Pontederia cordata*.

Turning now to the reproductive behavior which has led to the development of societies we find a most extraordinary parallelism between the group of bees as a whole and that of the wasps as described in my previous lecture. The progress from the solitary condition, shown in more than 95 per cent. of the species, to the conditions in the most

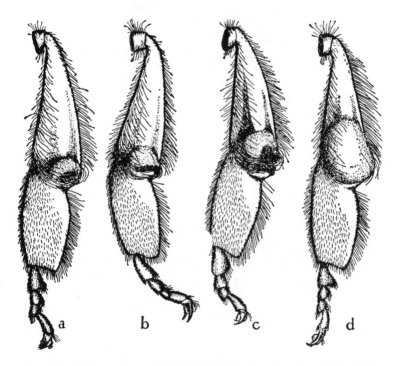

Fig. 37. — Outer surfaces of the left hind leg of worker bees in successive stages of pollen accumulation. *a*, from a bee just beginning to collect. The pollen mass lies at the entrance of the basket. The planta is extended, thus lowering the auricle. *b*, slightly later stage, showing increase in pollen. The planta is flexed, raising the auricle. The hairs extending outward and upward from the lateral edge of the auricle press upon the lower and outer surface of the small pollen mass, retaining and guiding it upward into the basket. *c* and *d*, slightly later stages in the successive processes by which additional pollen enters the basket. (After D. B. Casteel.)

highly socialized form, the honey-bee, is, so to speak, a
repetition of the various wasp *motifs* set in a different key.
Every one of the thousands of species of solitary bees has
its own peculiarities of behavior, but the differences are

FIG. 38. — Pollen manipulation of honey-bee. *A*. Flying bee, showing
manner of manipulating the pollen with the fore and middle legs. The
fore legs are removing the pollen from mouthparts and face; the right
middle leg is transferring the pollen on its brush to the pollen combs
of the left hind planta. A small amount of pollen has already been
placed in the baskets. *B*. Flying bee showing portion of middle legs
touching and patting down the pollen masses. *C*. Inner surface of hind
leg bearing a complete load of pollen. *a*. Scratches in pollen mass
caused by pressure of the long projecting hairs of the basket upon the
pollen mass as it has been pushed up from below. *b*. Groove in the
pollen mass made by the strokes of the auricle as the mass projects
outward and backward from the basket. (After D. B. Casteel.)

usually so insignificant that the habits as a whole are very
monotonous.   With the exception of the parasitic bees,
which have been secondarily evolved from non-parasitic
forms, all the solitary bees make their nests either in the
ground or in the cavities of plants, in crevices of walls,
etc., or construct earthen or resin cells (Fig. 39).   Some
species line their nest cavities with pieces of the leaves or

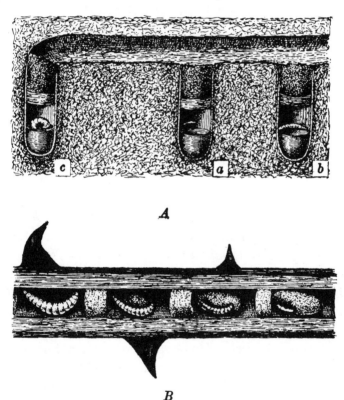

FIG. 39. — Nests of solitary bees. *A.* Nest of *Colletes succinctus*
in the ground. After Valery Mayet. *a,* cell provisioned and sup-
plied with an egg; *b,* cell with young larva; *c,* with older larva.
B. Nest of a small carpenter bee (*Ceratina curcurbitacea*) in a
hollow Rubus stem; showing egg, three larvæ of different stages
and bee-bread in three of the cells. (After Dufour and Perris.)

petals of plants, with plant-hairs or particles of wood, or with films of secretion which resemble celluloid or gold-beater's skin. Most of these materials, as will be noticed, are derived from plants. The nest usually consists of sev-eral cylindrical or elliptical cells arranged in a linear series or more rarely in a compact cluster, and as soon as a cell has been completed, it is provisioned with a ball or loaf-shaped mass of pollen soaked with honey and called " bee-bread," an egg is laid on its surface and the cell is closed. We have here again the typical mass provisioning of the solitary wasps, very similar to that of the Eumeninæ, except that vegetable instead of animal substances are provided for the young. Nevertheless, the pollen and honey are ideal foods, since the former is rich in proteins and oils and the latter in sugar and water, and both contain sufficient amounts of various salts for the growth of the larvæ. As in the case of the solitary wasps the mother bee dies before her prog-eny emerge.

Just as among the solitary wasps, we often find female solitary bees nesting in close association with one another, and in some species (*Halictus longulus, Panurgus, Euglossa, Osmia vulpecula* and *parietina, Eucera longicornis*) the females, though occupying separate nests, nevertheless build a common entrance tunnel. Still there is nothing in these arrangements to indicate that they could lead to the forma-tion of true societies. There are, however, a few cases which might be regarded as sub-social, since the mother bee survives the development of her progeny and shows more interest in their welfare than is implied by the mere mass provisioning of the cells. Two such cases are repre-sented by the European *Halictus quadricinctus*, observed by Verhoeff, and *H. sexcinctus,* observed by Verhoeff, von Buttel-Reepen and Friese. The female of the former bee digs a long vertical tunnel in the ground and at its lower end a chamber in which she constructs a number of earthen cells, arranged in the form of a rude comb. These cells, of which there may be as many as 16 to 20, are successively provisioned and closed, but the mother is long-lived, guards the nest and may even survive till the young emerge. Hence there is here an actual though apparently very brief contact of the mother with her adult offspring.

Certain peculiarities in the life-history of Halictus may be conceived to tend still further towards social development. According to our present unsatisfactory knowledge of these bees, at least some of the species have two annual generations. The spring generation consists of fecundated females that have over-wintered from the previous fall. These give rise to a summer generation consisting entirely of females. Their eggs develop parthenogenetically, but produce both males and females, forming the fall generation. The males soon die, but the fecundated females go into hibernation. As von Buttel-Reepen suggests, a society might be readily established in a form like *H. quadricinctus* if the parthenogenetic generation of females were to remain with their mother and extend the parental nest. This would be essentially what we find in the lower social wasps like Polistes.[5]

A still more interesting case has been found by Dr. Hans Brauns among the bees of the genus Allodape which belong to the gastrilegous division and are closely related to our small carpenter bees of the genus Ceratina, so abundant in hollow stems of the elder and sumach. Dr. Brauns made his observation in South Africa, where he has been living for many years, and kindly sends me the following unpublished data for use in this lecture:

" The species of Allodape nest in the dry, hollow stems

of plants, very rarely in galleries in the soil. In both cases they gnaw out cavities or occupy those already in existence. Plant stems with pithy contents, like those of Rubus, Liliaceæ, Aloë, Amaryllidaceæ, Asparagus, Acacia thorns, etc., are preferred. Three different groups of species may be distinguished according to the method employed in provisioning the young. These three groups may also prove to be useful as morphological sections of the genus, since the majority of Allodape species, especially the smaller ones, are very difficult to distinguish in the female sex. The males yield better characters, though there are few plastic characters in the genus. Most of the descriptions drawn from single captured specimens have little value. Fanatical describers, like some of your countrymen, merely make the work of the monographer more difficult or more unattractive or even well nigh impossible in a genus which is almost as monotonous as Halictus. The three different methods of provisioning which I have been able to establish are the following:

" 1. The most primitive species, observed only on a few occasions. The mother bee collects in the nest tube as much bee-bread in single loaves or packets as the larva will require up to the time of pupation, precisely as in other solitary bees, *e.g.*, as in Ceratina, the form most closely related to Allodape. The single food-packets are arranged one above the other in the hollow stem and each is provided with an egg. The larva holds itself to the food-packet by means of peculiar, long, segmental appendages, which I have called provisionally ' pseudopodia,' and consumes its single packet till it is time for pupation. The size of the packet corresponds to the size of the particular species, much as in Ceratina, and each packet nourishes only a single larva. The latter holds its appendages spread

out like those of a spider and is closely attached to the
packet like the larvæ of such solitary bees as Ceratina. So
far there is no departure from the conditions in the solitary
Apidæ. There is, however, one fundamental difference:
Whereas Ceratina after provisioning and oviposition closes
off each cell with a partition of gnawed plant materials, and
therefore makes a series of individual cells, Allodape *constructs absolutely no partitions.* The food-packets, each
large enough for a *single larva* and each furnished with a
single egg, though arranged in a linear series one behind the
other in the nest tube, as in Ceratina, Osmia, etc., lie freely
one on top of the other and are not separated by partitions
of the materials above mentioned. The lowermost packet
is the oldest and is therefore usually found to bear a larva
while each of the upper packets bears an egg. This difference, as you will admit, must be regarded as of fundamental importance. In these more primitive species the
mother does not come into contact with the larva since the
latter has been provided *once for all* with sufficient food to
last it till it pupates, precisely as in the solitary bees and
wasps. The pseudopodia can not therefore have the function of exudate organs but merely serve to attach the larva
mechanically to the food-packet. This transition from
isolated cells to a simple unseparated series of packets is,
of course, very interesting and significant.

" 2. Rather common, small and medium-sized species.
The mother bee glues a number of eggs, each by one pole
and in a *half spiral* row, determined by the curvature of the
tubular cavity, to the wall of the nest, usually near the
middle, *i.e.*, a little above or a little below. One common
species I have also seen occupying tubular cavities in the
earth with a similar arrangement of the eggs. The hatching
larvæ hold fast to the walls of the tube by means of their

pseudopodia and *are all at the same level with their heads
directed towards the entrance to the cavity.* From time to
time the mother brings in a small lump of bee-bread and
deposits it in the midst of the hungry heads. The larvæ
therefore all eat *simultaneously of the same mass of bee-
bread.* During their last moult the mature larvæ lose the
pseudopodia and become pupæ, which come to lie one
behind the other in the tubular nest cavity. In these species, therefore, the mother remains in continuous contact
with the larvæ.

" 3. The majority of species, from those of small to those
of the largest size. The mother bee lays her eggs singly
and loosely on the bottom of the nest tube. In proportion
to the size of the bee the eggs are very, one might say abnormally, large and seem to be laid at longer intervals.
The mother bee feeds the individual larva, which *clasps
the particle of bee-bread* with its two large pseudopodia
so that that it *has the food all to itself.* When a nest that
has been occupied for some time by a mother bee is examined, one or several larvæ, each with its own pellet of bee-
bread, are found in the position which I have described.
Later the daughters help their mother in provisioning the
larvæ. When the colony has become populous the cavity
of the tube is found to be stuffed with larvæ and pupæ in
all stages. The latest egg, however, almost always lies on
the floor of the tube. And since the mother bees must
always go to the bottom to feed the youngest larvæ, the
contents of the tube are often intermingled, though the
larger larvæ and the pupæ are mostly nearer the opening
and therefore uppermost. In these species, also, the larvæ
lose the pseudopodia during the last moult."

Brauns' observation on Allodape are of great interest
and importance because they reveal within the limits of a

single genus a series of stages beginning with a mass-provisioning of the young, like that of the solitary bees and wasps, and ending with a stage of progressive provisioning. And not only has the latter led to an acquaintance of the mother with her offspring but in the third group of species described by Brauns to an affiliation of the offspring with the mother to form a cooperative family or society. It would seem that this condition must have had its inception, as Brauns suggests, in so simple a matter as the omission of the series of partitions which all other solitary bees construct between their provisioned cells. The final stage in which the individual larvæ are fed from day to day by the mother and her daughters with small pellets of food is not essentially different from what we shall find in the bumble-bees and certain ants.

Yet these rudimentary societies of certain species of Halictus and Allodape must not be regarded as the actual precursors or sources of the conditions which we observe in the three groups of social bees, namely, the Bombinæ, or bumble bees, the Meliponinæ, or stingless bees, and the Apinæ, or honey-bees. Though these all belong to the podilegous division, no one has been able to point out their putative ancestors among existing solitary bees, and it is evident that we can neither derive them from one another nor from any single known extinct genus. Each possesses its own striking peculiarities and each is an independent branch from the ancestral stem now vaguely represented by the solitary bees. The bumble-bees are the most primitive, the honey-bees the most specialized, while the stingless bees exhibit a combination of primitive and specialized characters different from those of either of the other subfamilies. But just as all the social wasps differ from the solitary wasps in employing a peculiar nest mate-

rial — paper — so the three groups of social bees differ
from the solitary bees in using another peculiar nest material — wax. This material is, however, a true secretion,
which arises in the form of small flakes from simple glands
situated between the abdominal segments of the insects
(Fig. 40). The three groups of social bees also agree in

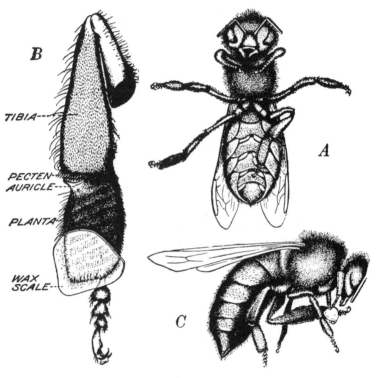

Fig. 40. — *A*. Ventral view of worker honey-bee in the act of removing
a wax-scale. *B*. Inner surface of left hind leg, showing the position of
a wax-scale immediately after it has been removed from the wax pocket.
The scale has been pierced by seven of the spines of the pollen combs
of the first tarsal segment of the planta. *C*. Side view of a worker bee
showing position of wax-scale just before it is grasped by the fore legs
and mandibles. The scale is still adhering to the spines of the pollen
combs. The bee is supported upon the two middle legs and a hind leg
as in *A*. (After D. B. Casteel.)

the structure of the hind tibia, the outer surface of which is not only broadened as in solitary forms but smooth and shining with recurved bristles along the edges (Fig. 36). This is called the corbicula and among solitary bees is known to occur only in Euglossa.

The bumble-bees represent a stage of societal development of the greatest interest to the evolutionist. Of these large insects about 200 species are known, mostly confined to Eurasia and North America. They prefer rather cool climates and several species occur in the arctic regions or at high elevations. Their habits have been carefully studied by several European entomologists, notably by Hoffer, Wagner, Lie-Petersen and Sladen, and are beginning to attract students in this country. We know very little about the species of Central and South America and the East Indies.[6]

In temperate regions bumble-bee colonies are annual developments, like those of our northern species of Vespa and Polistes. The large fecundated female or queen overwinters precisely like the females of the solitary wasps and starts her colony in the spring. She chooses some small cavity in the ground or in a log, preferably an abandoned mouse-nest, and after lining it with pieces of grass or moss or rearranging the pieces already present, proceeds to the important business of establishing her brood. The various stages in this behavior have been carefully observed by Sladen: " In the center of the floor of this cavity she forms a small lump of pollen-paste, consisting of pellets made of pollen moistened with honey that she has collected on the shanks (tibiæ) of her hind legs (Fig. 41 a). These she molds with her jaws into a compact mass, fastening it to the floor. Upon the top of this lump she builds with her jaws a circular wall of wax, and in the little cell so formed she lays her first batch of eggs (Fig. 42 *Ba*), sealing it over with wax by closing in the top of the wall with her jaws as soon as the eggs have been laid. The whole structure is about the size of a pea. . . . The queen now sits on her eggs day and night to keep them warm, only leaving them to collect food when necessary. In order to maintain animation and heat through the night and in bad weather when food cannot be obtained, it is necessary for her to lay in a store of honey. She therefore sets to work to construct a large waxen pot to hold the honey (Fig. 41 *b*, 43,

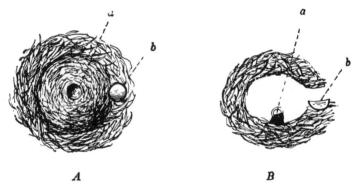

A   B

Fig. 41.— Incipient nest of bumble-bee. *a*, Pollen and first eggs;
b, Honey pot. (After F. W. L. Sladen.)

44). This pot is built in the entrance passage of the nest, just before it opens into the cavity containing the pollen and eggs, and is consequently detached from it. The completed honey pot is large and approximately globular, and is capable of holding nearly a thimbleful of honey."

Up to this point the behavior of the queen is much like that of the solitary bee which makes and closes her cell after providing it with provisions and an egg, but a significant change now supervenes. The eggs hatch after about four days and the further events are described by Sladen as follows: " The larvæ devour the pollen which forms their bed, and also fresh pollen which is added and plastered onto the lump by the queen. The queen also feeds them with a liquid mixture of honey and pollen, which she prepares by swallowing some honey and then returning it to her mouth to be mixed with pollen, which she nibbles from the lump and chews in her mandibles, the mixture being swallowed and churned in the honey-sac. To feed the larvæ the queen

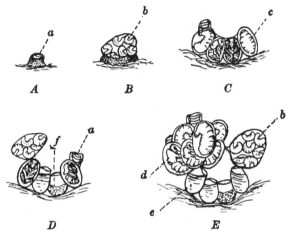

Fig. 42. — A to E. Diagrams of successive stages in the development of the bumble-bee's brood. a, eggs; b, young larvæ; c, full grown larva; d, pupa; e, old cocoon used as a honey-pot; f, old cocoon used as a pollen pot. (After F. W. L. Sladen.)

makes a small hole with her mandibles in the skin of wax that covers them, and injects through her mouth a little of the mixture among the larvæ which devour it greedily. Her abdomen contracts suddenly as she injects the food, and as soon as she has given it she rapidly closes up the hole with the mandibles. While the larvæ remain small they are fed collectively, but when they grow large each one receives a separate injection."

Here we have a beautiful transition from mass to progressive provisioning. Sladen then describes the further development of the brood: " As the larvæ grow the queen adds wax to their covering, so that they remain hidden (Fig. 42 *BEb*). When they are about five days old the lump containing them, which has hitherto been expanding slowly, begins to enlarge rapidly, and swellings, indicating

Fig. 43.— Incipient nest of *Bombus terrestris,* showing honey-pot and mass of wax enclosing young brood and grooved for the accommodation of the body of the queen while incubating. (After F. W. L. Sladen.)

the position of each larva, begin to appear in it. Two days later, that is, on the eleventh day after the eggs were laid, the larvæ are full-grown, and each one then spins around itself an oval cocoon, which is thin and papery but tough (Fig. 42 *Cc*). The queen now clears away most of the brown wax covering, revealing the cocoons, which are pale yellow. These first cocoons number from seven to sixteen, according to the species and the prolificness of the queen. They are not piled one on another, but stand side by side, and they adhere to one another very closely, so that they seem welded into a compact mass. They do not, however, form a flat-topped cluster, but the cocoons at the sides are

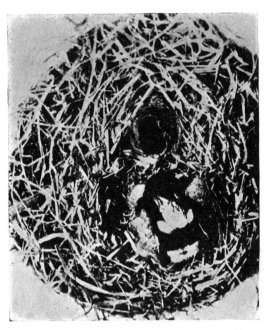

Fig. 44. — Same as Fig. 43, showing the queen *Bombus terrestris* lying in the groove and incubating the young brood. (After F. W. L. Sladen.)

higher than those in the middle, so that a groove is formed; this groove is curved downwards at its ends (Fig. 43), and in it the queen sits, pressing her body close to the cocoons and stretching her abdomen to about double its usual length so that it will cover as many cocoons as possible;

at the same time her outstretched legs clasp the raised
cocoons at the sides (Fig. 44). In this attitude she now
spends most of her time, sometimes remaining for half-an-
hour or more almost motionless save for the rhythmic expansion and contraction of her enormously distended abdomen, for nothing is now needed but continual warmth to
bring out her first brood of workers. In every nest that I
have examined the direction of the groove is from the
entrance or honey-pot to the back of the nest, never from
side to side. By means of this arrangement the queen,
sitting in her groove facing the honey-pot — this seems to
be her favorite position, though sometimes she reverses it
— is able to sip her honey without turning her body, and at
the same time she is in an excellent position for guarding
the entrance from intruders."

The eggs laid by the queen during the early part of the
summer are fertilized and therefore produce females, but
the larvæ, owing to the peculiar way they are reared,
secure unequal quantities of nutriment and therefore vary
considerably in size, though they are all smaller than their
mother. Individuals scarcely larger than house-flies are
sometimes produced, especially in very young colonies. All
of these individuals have been called workers, although
they have essentially the same structure as the queen.
They are assisted in emerging from their cocoons by their
mother or sisters and forthwith take up the work of collecting pollen and nectar and of enlarging the colony. The
queen now remains in the nest and devotes herself to laying eggs, while the nest is protected, new cells are built and
the additional broods of larvæ are fed by the workers.
They also construct honey-pots and special receptacles for
pollen or store these substances in cocoons from which
workers have emerged (Fig. 45). Later eggs are also laid

by the workers but being unfertilized develop into males.
As the colony grows and becomes more prosperous, some of
the larvæ derived from fertilized eggs laid by the queen
are abundantly fed and develop into queens.   Like the
queens of the social wasps, these do not emerge from their
cocoons till the late summer, and like the queen wasps, they

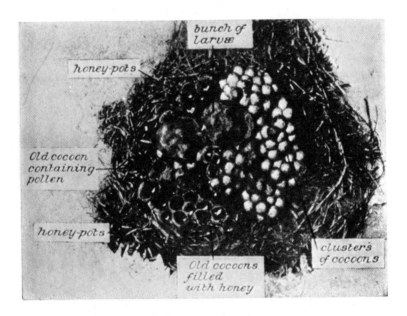

FIG. 45. — Comb of *Bombus lapidarius* showing clusters of worker
cocoons, masses of enclosed larvae, half-full honey-pots and pollen-pot.
(After F. W. L. Sladen.)

disperse, after mating with the males, and alone of all the
colony survive the winter to start new colonies the following spring.   In South America, where, according to von
Ihering, bumble-bee colonies are perennial, new nests are
formed by swarming as among the social wasps of the same
region.   Bumble-bee colonies are, as a rule, not very populous, 500 individuals constituting an unusually large so-

ciety. In many cases there are scarcely more than 100 to 200.

I have called attention to the fact that the workers are precisely like the queens, or fertile females, except that they have been more or less inadequately fed during the larval stages and are therefore smaller. They are the result of a high reproductive activity on the part of the queen under unfavorable trophic conditions that do not permit the offspring to attain their full stature. In certain species that live permanently under even more unfavorable conditions, like those in the arctic regions, the worker caste is completely or almost completely surpressed. During 20 years of residence in Tromsö, Norway, Sparre Schneider failed to find a single worker of *Bombus kirbyellus,* and those of *B. hyperboreus* were extremely rare. Probably the queens of these species are able to rear only a few offspring and these are all or nearly all males and queens, though, during the short arctic summer, at least in Finland and Lapland, the mother insects work late into the nights. But the worker caste may also disappear as a result of the opposite conditions, that is, an abundance of food. We found this to be the case with the workerless parasitic wasps, *Vespa arctica* and *austriaca.* In north temperate regions the genus Bombus has given rise to a number of parasitic species, which have been included in a separate genus, Psithyrus. These bees are very much like Bombus, in the nests of which they live, but just as in the two species of Vespa and for the same reasons, their worker caste has been suppressed.

The foregoing account shows that the bumble-bees are very primitive and represent an interesting transition from the solitary to the social forms, since the queen while establishing her colony behaves at first like a solitary bee but later gradually passes over to a stage of progressive provisioning and affiliation of her offspring and thus forms a true society. The cells are also essentially like those of solitary bees, except that they are made of wax, but even in the secretion of the wax the bumble-bees represent the primitive conditions, as compared with the stingless bees and honey-bees, since the substance is exuded between both the dorsal and ventral segments of the abdomen.

Part 2

The Meliponinæ, or stingless bees, are a very peculiar group of nearly 250 species, all confined to warm countries. Fully four fifths of the species occur only in the American tropics and only about one fifth in the Ethiopian, Indomalayan and Australian regions. All the Old World and the majority of the Neotropical forms belong to the genus Trigona; the remainder of the American species are placed in a separate genus, Melipona. The stingless bees are much less hairy and much smaller than the bumble-bees. Some of the species of Trigona measure less than 3 mm. in length and are therefore among the smallest of bees. The colonies vary greatly in population in different species. According to H. von Ihering, those of Melipona may comprise from 500 to 4,000, those of Trigona from 300 to 80,000 individuals. The name stingless as applied to these insects is not strictly accurate, because a vestigial sting is present. It is useless for defence, however, so that many of the species are quite harmless and are called " angelitos " by the Latin-Americans. But some forms are anything but little angels. When disturbed they swarm at the intruder, bury themselves in his hair, eye-brows and beard, if he has one, and buzz about with a peculiarly annoying, twisting movement. Others prefer to fly into the eyes, ears, and nostrils, others have a *penchant* for crawling over the face and hands and feeding on the perspiration, or bite unpleasantly, and a few species spread a caustic secretion over the skin. On one occasion in Guatemala large patches of epidermis were thus burned off from my face by a small swarm of *Trigona flaveola.*[7]

There are three morphologically distinct castes. The queen differs from the worker in the smaller head, much more voluminous abdomen, more abundant pilosity, and in the form of the hind legs, the tibiæ of which are reduced in width and furnished with bristles also on their external surfaces, while the metatarsi are elongate, rounded and apically narrowed. The worker, therefore, really represents the typical female of the species morphologically, except that she is sterile, whereas the queen, except in her ovaries, exhibits a degeneration of the typical secondary characters of her sex. There is only one mother queen in a colony, but a number of young daughter queens are tolerated. New colonies are formed by swarming, that is, by single young queens leaving the colony from time to time, accompanied by detachments of workers, to found new nests. The body of the old queen is so obese and heavy with eggs and her wings are so weak that she can not leave the nest after it is once established.

The nests of the Meliponinæ are extremely diverse in structure. They are usually in hollow tree-trunks or branches, less frequently in walls. Some of the species nest in the ground, and a few (*T. kohli, fulviventris, crassipes, etc.,*) actually build in the centers of termite nests. The nest is made of wax, which most of the species mix with earth, resin or other substances, so that it is chocolate brown or black and is called " cerumen." The wax is secreted only between the dorsal segments of the abdomen, and is produced by the males as well as by the workers — the one case in which a male Hymenopteran seems to perform a useful social function. The workers not only collect nectar and pollen but they seem to have a greater propensity than other social bees for gathering propolis, resins

Fig. 46. — Cerumen spout, or nest entrance of a large colony of *Trigona heideri* Friese nesting in a hollow tree at Kartabo, British Guiana. About natural size. (Photograph by Mr. John Tee Van.)

and all kinds of gums and sticky plant-exudations. And unlike other bees they are also fond of visiting offal and the feces of animals. One species is said even to eat meat (*T. argentata,* according to Ducke).

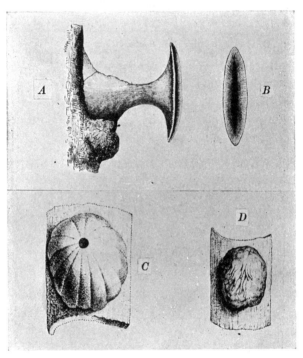

Fig. 47. — Cerumen entrances to nests of Meliponine bees. *A,* of *Trigona læviceps* of India in profile; *B,* same seen from the front (After C. S. P. Parish); *C,* nest entrance of *Melipona quinque-fasciata; D,* nest entrance of *Trigona limao.* (After F. Silvestri.)

The entrance to the nest may be a simple hole, but more often it is a projecting cerumen spout or funnel, which differs considerably in different species (Fig. 46). In some East Indian Trigonas its lips are kept covered with sticky propolis to prevent the ingress of ants and other intruders (Fig. 47 *AB*). In most of the South American species its

orifice is guarded during the day by a special detachment
of workers and is closed at night with a cerumen plate or
screen. The interior of the nest presents a peculiar appearance. If it is in a hollow tree-trunk or branch the cavity
is closed off at each end by a thick lump or plate of cerumen
(the "batumen"). The nest proper (Figs. 48 and 49),
constructed in the tubular space thus preempted, consists
of two parts, one for the brood and one for the storage of
various foods and building materials. The brood portion
consists typically of a hollow spheroidal envelope of irregu-
lar, interconnected cerumen laminæ, forming the walls of
an elaborate system of anastomosing passage-ways and
enclosing a large central space occupied by a series of
combs of hexagonal cells. There is only one layer of cells in
each comb and they all open upwards, not downwards as
in the social wasps. In some species the combs are regular
and disc- or ring-shaped structures, in others they are arranged in a spiral or more irregularly. Their cells are used
exclusively for rearing the brood. In Melipona and some
species of Trigona they are all of the same size, but in
several South American species of the latter genus single
larger cells are constructed, especially towards the periphery of the combs, for the rearing of queen larvæ. All this
elaborate arrangement would seem to be a preparation for a
very specialized system of caring for the young, but such is
not the case. The workers, precisely as if they were solitary
bees, put a quantity of pollen and honey into each cell, and
after the queen has laid an egg in it, provide it with a
waxen cover, so that the larva is reared exactly like that
of a solitary bee. There is mass but not progressive provisioning and the adult bees do not come in contact with
the growing larvæ. The queen-cells are treated in the same
manner, the only difference being that they are provided

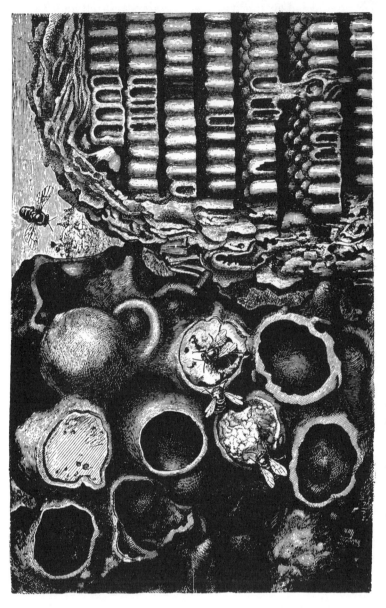

Fig. 48.—Portion of nest of *Melipona scutellaris*, showing brood-comb (to the right) and the large honey-pots and pollen-pots (to the left). (Subdiagrammatic drawing from Emile Blanchard.)

with a greater amount of pollen and honey. Among the species whose queen cells are no larger than those in which the workers and males are reared, the queens emerge with small ovaries and develop them later, but among the spe-

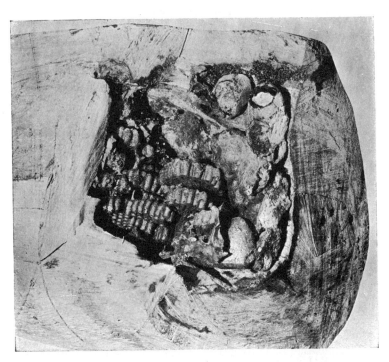

Fig. 49. — Nest of *Melipona* sp. in hollow log, showing brood-comb (to the left), pollen-pots and honey-pots (to the right). (Photograph by Dr. E. F. Phillips. About one half natural size.)

cies which build large cells for the queens, the latter emerge with the ovaries fully developed. These differences are of considerable interest in connection with the queen honey-bee.

Outside the cerumen involucre enclosing the brood combs the workers construct large elliptical or spherical pots, some for the storage of honey, others for pollen and in some species still others for propolis. In one species (*T. silvestrii*) the pollen-pots are long and cylindrical while the honey-pots are small and spherical. This same arrangement is known to occur also in one of the European bumble-bees (*B. pomorum*). The Meliponinæ may also store other substances in the outer portions of their nests. In one nest of a black Trigona, which I observed at Kalacoon, in British Guiana, there were several lumps, each weighing 10 to 20 grams, of a hard substance closely resembling sealing wax in color and consistency. The species which build their nests freely on the branches of trees cover them with protective layers of cerumen arranged like those surrounding the brood combs.

The Old World Trigonas (*canifrons*, according to W. A. Schulz) and some of the South American species (*timida*, *silvestrii* and *cilipes*, according to Silvestri; *silvestrii* and *muelleri*, according to H. von Ihering; and a very small undescribed species allied to *T. goeldiana*, which I found in British Guiana) represent a more primitive stage in the construction of the nest (Fig. 50). The brood cells in these forms are elliptical and are not arranged in a comb but are isolated or loosely connected with one another by delicate waxen beams, or trabeculæ. They are therefore essentially like the cells of bumble-bees and solitary bees. It should be noted also that the Meliponinæ, like the bumble-bees, tear down their cells after they have been used and construct new ones in their places.

The rearing of the brood of all the castes in closed cells, after the manner of the solitary bees, is very significant, since it is the only case among the social Hymenoptera of a complete lack of contact between the adults and the larvæ. Even the bumble-bees open their cells from time to time and feed the older larvæ, and among the honey-bees the cells remain open throughout larval development. It is obvious that the Meliponinæ have either retained unaltered the ancient method of rearing the young in closed cells, employed by all the solitary bees, or have reverted to it

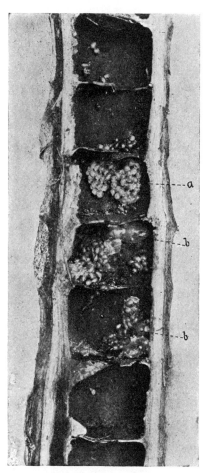

Fig. 50.—Nest of a small Trigona (worker only 2.5 mm. long!) representing a new species near *T. goeldiana* Friese, in the hollow internodes of a small *Cecropia angulata* Bailey at Kartabo, British Guiana. *a*, brood-cells containing adult larvæ and pupæ; *b*, honey-pots. Slightly enlarged. (Photographed by Mr. John Tee Van.)

after practicing a method more like that of the bumble-bees or honey-bees. As there seem to be no cogent reasons for adopting the latter alternative, I am inclined to believe that the former is the more probable and that unlike the wasps these highly social bees have never passed through a stage of actual trophallactic contact between mother and offspring. After considering the honey-bees I shall return to this question.

The Apinæ, or honey-bees, are separated by a wide gap from the Meliponinæ and Bombinæ and their origin is still wrapped in obscurity. Certain species, referred by their authors to the genus Apis, are recorded from the European Miocene (*A. adamitica*), Baltic Amber and Upper Oligocene (*A. meliponoides* and *henshawi*).[8] The genus as it exists to-day comprises only four species: *dorsata, florea, indica* and *mellifica*, the common honey-bee of our apiaries. *A. indica* and *mellifica* are so very similar in structure and habits and hybridize so readily that both Friese and von Buttel-Reepen have regarded the former as a mere race, or sub-species of the latter. Von Buttel-Reepen, however, has recently raised *indica* to specific rank on what seem to me to be rather dubious grounds.[9] Inasmuch as *dorsata, florea* and *indica* are confined to the Indomalayan region, it has been usually assumed that *mellifica*, though now cosmopolitan, is also of South Asiatic origin. But von Buttel-Reepen believes that it had its origin in Germany and bases his opinion on the existence of the above-mentioned fossils in Germany and on the fact that the true *mellifica* did not exist in India till it was introduced by Europeans. The bee originally kept in that country by the natives was *indica*. The eminent melittologist's view is so startling that one is tempted to suppose that he, like some other German investigators, is the victim of a desire to make his fatherland the source of all good things. The following considerations seem to me to leave little ground for his opinion: First, if *mellifica* was not originally present in India it is probably because *indica* happens to be the South Asiatic race of the species. Second, the type of *mellifica,* that is, the form to which Linnaeus first gave the name, is of course, the dark German, or northern race, and it is natural to regard the many local races and varieties in other parts of Europe, in Africa and Asia as mere modifications of the German type. But such a procedure is unwarranted in phylogeny, since the selection of the German race as the specific type was a mere taxonomic accident. Had a Hindoo entomologist preceded Linnaeus, *indica* would be the type of the species, and the Hindoo, aware of the existence of two other species of Apis in his and neighboring countries and nowhere else, would properly regard the genus as of South Asiatic origin and the species *indica* as having spread to Europe and Africa and as having produced among others a dark Germanic race. Third, it is by no means certain that the fossil forms, which have been described from imperfect specimens, really belong to Apis or that they are in the direct line of descent to the genus. And even if we admit this to be the case, it does not follow that they must have originated in Germany. Fourth, granting that a race of *mellifica* existed in that country during the Miocene, it must either have become extinct during the Ice Age or have been driven into Southern Europe. That this identical race and not a new one arising from southern forms later returned to Germany is a pure supposition. Fifth, the tropical origin of the honey-bee is indicated by its inability to form new colonies except by swarming, precisely like the tropical Meliponinæ and Vespidæ. And while it is true that the climate of Central Europe during the Oligocene and Miocene was tropical or subtropical, the existence at the present time of at least three distinct species of Apis in the Indomalayan region and nowhere else makes it seem much more probable that the ancestors of *mellifica* emigrated from Asia into Europe than in the reverse direction, especially as Southern Asia is a well-known center from which many other animals have been distributed. The spread of the honey-bee throughout the world is evi-

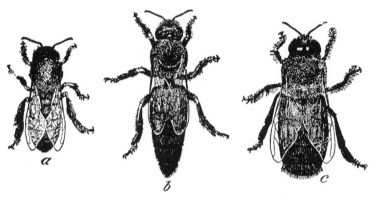

Fig. 51.— Honey-bee (*Apis mellifica*); *a*, worker; *b*, queen; *c*, male (drone). Twice natural size. (After E. F. Phillips.)

dently due to its extraordinary adaptability to the most diverse flowers and to a great range of temperature, to its habit of storing large quantities of honey and its ability to maintain a rather high temperature in the hive during periods of cold weather. This unusual plasticity is peculiar to the species and is not the result of domestication. The insect, in fact, has never been domesticated.

Like the Meliponinæ the species of Apis have three well-developed castes (Fig. 51). Normally there is only a single queen in the colony and she will not tolerate the presence even of another young queen. Swarming takes place by the old queen leaving the colony accompanied by a large detachment of workers when a young queen is about to emerge from her cell, and if several young queens are to emerge in succession, the older leaves before the next appears. It will be noticed that this is different from the swarming of the Meliponinæ, since in these bees the old queen remains in the nest and the young queens accompany the swarms of workers. When the queen's eggs are fertilized they develop into workers or queens according to the way the larvæ are fed, but when unfertilized, into males or drones, as is also the case with the eggs that are sometimes laid by workers. All the species of Apis make pendent combs of pure wax, which is secreted only by the workers and only between the ventral segments of the abdomen. The combs differ from those of the Meliponinæ and wasps in consisting of two layers of hexagonal cells, and the brood cells remain open throughout larval development, the young being fed progressively. The three species of Apis represent as many different phylogenetic stages, which may be briefly described.

A. dorsata is the largest and most primitive form (worker 16-18 mm.; queen 23 mm.; drone 15-16 mm.). It builds from the lower surface of a branch a single large semicircular comb, sometimes with a superficial area of a square meter. Its cells are regularly hexagonal and all alike. Part of them are used for storing honey, the remainder for the brood. In this species, therefore, the workers, queens and drones are all reared in cells of the same size and shape like the species of Melipona and many Trigonas. *A. dorsata* is a nomadic bee which builds its comb where flowers are abundant and after they have ceased to bloom deserts the structure and builds a new one in fresh pastures. Owing to this habit all the attempts that have been made both in Europe and the United States to establish this bee in apiaries have failed.

A. florea is the smallest species in the genus (worker
7-8 mm.; queen 13-14 mm.; drone 12 mm.) and in certain
respects represents an interesting transition between
dorsata and *mellifica*. The drone is peculiar in having a
finger-shaped process on the inner border of the hind meta-
tarsus. Like *dorsata, florea* makes a single pendent comb,
but it is much smaller and narrower and consists of four
different kinds of cells. At the base where the comb sur-

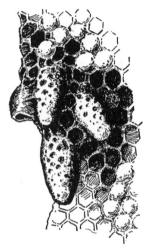

FIG. 52. — Queen-cells of the honey-bee.   Natural size.   (After
E. F. Phillips.)

rounds the supporting branch the cells are hexagonal,
large and deep and are used for storing honey. Below
these there is a broad zone of small hexagonal cells for the
worker brood and at the apical fourth still larger hexagonal cells for the drone brood. Finally, attached to the free
border and depending like stalactites there are several
long conical cells for the queen larvæ.

A. mellifica and its subspecies *indica*, etc., differ from
dorsata and *florea* in nesting in hollow cavities (tree-trunks,

caverns, hives) and in constructing several pendent combs
side by side, each presenting the types of cells seen in *florea*,
except that there is no special type for storage, the honey
being kept in cells like those used for rearing the worker
brood. Moreover, the queen cells are never built on drone
but only on worker comb.[10]   The singular shape of the
queen cells (Fig. 52), so very different from the hexagonal

Fig. 53. — Comb built by a colony of honey-bees on the branches of
a tree. From a specimen in the American Museum of Natural History
by which this photograph and Fig. 54 were contributed.

cells, and the fact that they are the only cells torn down
by the workers after being used, indicate that they are
archaic structures of considerable phylogenetic significance.
They are, indeed, reminiscent of the only type of cell constructed by the bumble-bees. But owing to the fact that
conical queen cells occur only in *A. florea* and *mellifica* and
are the same in both species we are unable to advance any
reasons for their retention among cells of the highly spe-

FIG. 54. — Same comb as shown in Fig. 53, seen from the end.

cialized hexagonal type. On rare occasions a swarm of
honey-bees, failing to find a hive or hollow tree-trunk, will
construct its comb among the branches of trees or bushes.

Such exposed nests have been described by Bouvier, and there is an unusually fine example in the American Museum of Natural History (Figs. 53 and 54). It will be noticed that in form each comb resembles the single comb of *A. dorsata* or *florea*.[11]

Except in the development of her ovaries, the queen honey-bee is a degenerate female, a mere egg-laying machine, entirely dependent on her worker progeny. The pollen-collecting apparatus of the hind legs, so well developed in the worker and so characteristic of the females of all non-parasitic, podilegous bees and of the queens of the bumble-bees, is undeveloped, her tongue and sting are shortened, her brain is smaller and she lacks the pharyngeal salivary glands of the worker. That these differences are due to larval feeding is proved by the experiment of transferring eggs and very young larvæ from worker to queen cells and *vice versa*. Transferring eggs from drone to worker or queen cells does not, however, alter the sex of the insect reared, since it develops from an unfertilized egg. Under normal conditions the time required for the development and the chemical composition of the food administered to the larvæ differ for the three castes. The difference in the rate of development is shown in the following table from Buttel-Reepen:

DURATION OF DEVELOPMENT IN THE HONEY-BEE

(After von Buttel-Reepen)

Development of the Brood	Queen (Days)	Worker (Days)	Drone (Days)
Duration of egg (embryonic) development......	3	3	3
Duration of larval development...............	6	6	6
Duration of spinning and resting period..........	2	4	7
Change of pupa to imago......................	5	8	8
Total	16	21	24

The composition of the food, which, for the queens and the earliest stages of the workers and drones, is a secretion ("royal jelly") of the pharyngeal glands of the worker nurses, is shown in von Planta's table taken from the same author:

COMPOSITION OF LARVAL FOODS OF THE HONEY–BEE
(After von Planta)

Queen Larvæ	Drone Larvæ			Worker Larvæ		
Percentages of Dried Substance (Average)	Under 4 Days	Over 4 Days	Average	Under 4 Days	Over 4 Days	Average
Proteid 43.14	55.91	31.67	43.79	53.38	27.87	40.62
Fat 13.55	11.90	4.74	8.32	8.38	3.69	6.03
Sugar 20.39	9.57	38.49	24.03	18.09	44.93	31.51

We see from these tables that the queen, although the largest of the three castes, reaches maturity in about 16 days. She is fed only on "royal jelly," without admixture of honey or pollen. These highly nutritious rations (43.14 per cent. protein) are undoubtedly responsible for her very rapid growth. The worker is given pollen and honey after the fourth day and requires 21 days to complete her development. The feeding of the drone is similar, but he receives less sugar and more fat and his development is protracted to 24 days.

The foregoing considerations suffice to show the complexity of the whole matter of sex-determination and caste-differentiation in the honey-bee. There are libraries of contentious discussion of these subjects, which can not, of course, be adequately treated in a general lecture.[12] Although all competent authorities agree that the drones arise from parthenogenetic or unfertilized eggs and the queens and workers from fertilized eggs, it is impossible, in the present state of our knowledge, to decide between two different theoretical interpretations of the facts. According to one, first stated by Dzierzon and more recently maintained by Weismann and his pupils and especially by Buttel-Reepen, the queen lays only one kind of egg, which is potentially indifferent but has its sex determined at the moment of fertilization; according to the other, advocated by Beard, von Lenhossek and Oscar Schulze, there are really two different kinds of eggs, one of which does not need to be fertilized and develops into the drone, whereas the other requires fertilization to be viable and develops into a queen or a worker. Normally the queen lays unfertilized eggs only in the large drone cells and fertilized eggs only in the small worker cells and the peculiar conical queen cells. The hypothesis that the difference in the width of the cells is the stimulus which causes the queen to close or open the duct of her spermatheca and thus prevent or permit the exit of sperm while the egg is passing from the ovaries on its way to being laid, can not be accepted, because the queen often oviposits in cells which have had only their basal portion completed. Moreover, this hypothesis will not apply to the Meliponinæ, which rear males and workers in closed cells of the same size, or to cases like those of the solitary wasps described in the preceding lecture and many solitary bees which regulate the size of the cell and the amount of provisions according to the sex of the offspring. These peculiar phenomena, first observed by Fabre in Osmia, Halictus and Chalicodoma, have been recently confirmed by Verhöff, Höppner and Armbruster.[13] The observations show that the female bee must be aware of differences between her eggs sometime before she begins to lay them, and certainly before there are any such stimuli as contact of her abdomen with the walls of the cell.

Referring to the Dzierzon theory, Phillips says:[14] " The facts observed in the apiary on which this belief is based are as follows: (1) If a queen is unable to fly out to mate or is prevented from mating in some way she usually dies, but if she does lay eggs, as she may, after three or four weeks, the eggs which develop are all males; (2) if when a queen becomes old her supply of spermatozoa is exhausted, her offspring are all males; (3) if a colony becomes queenless and remains so for a time, some of the workers may begin egg-laying and in this case too only males develop. The author has found that many eggs laid by drone-laying queens fail to hatch and, in fact, are often removed in a short time by the workers. This makes it impossible for us to accept Dzierzon's statement that all eggs laid by such a queen become males and the statement must be modified as follows: all of those eggs laid by a drone-laying queen which develop become males. The potentialities of the eggs which never hatch are not known. In addition to the facts here stated, the theory of the parthenogenetic development of the drone is supported by investigation of the phenomena of development in the egg." He becomes more explicit in the following passage: " If we take into consideration the important fact that not all eggs of an unfertilized (drone-laying) queen hatch, then the bee does not appear as an exception in nature. It seems clear, however, that the statement of Dzierzon that all the eggs in the ovary are male eggs can not be accepted and it is, in fact, not improbable that the eggs destined to be females die for want of fertilization, while the eggs destined to be males, not requiring fertilization, are capable of development. It should be understood that the casting of doubt on Dzierzon's theory of sex determination does not invalidate his theory in so far as it pertains to the development of males from unfertilized eggs."

There are also several cases of hybridization that seem

to indicate that Dzierzon's theory is only a partial or approximate interpretation. When a yellow Italian queen bee mates with a black German drone, the drone offspring, being fatherless, should of course be yellow like the mother, whereas the workers should combine the characters of both their parents. This is often the case, but although von Buttel-Reepen is a staunch supporter of Weismann and adheres rigidly to the Dzierzon theory, he is compelled to admit that occasionally "when an Italian queen is fecundated by a German drone, numerous blended hybrids ("Mischlinge") appear during the first year, but during the second year almost exclusively Italian, and in the third year exclusively Italian workers are produced, so that the population must be regarded as purely Italian." [15] He has himself witnessed this phenomenon and states that it has also been observed by Dönhoff, Dzierzon and Cori. The only explanation von Buttel-Reepen has to offer is that the sperm, stored in the spermatheca of the queen, may in the course of time be increasingly affected by the secretions of her spermophilous glands, which keep the paternal elements alive during the three to five years of her life. The facts can hardly be explained on Mendelian principles even if we make all due allowance for the impurity of the German and Italian strains that produced the hybrids. It looks as if, at least under certain conditions, workers might develop from unfertilized eggs. According to Onions,[16] the workers of a South African race of honey-bees are able to produce workers parthenogenetically, and Reichenbach, Mrs. A. B. Comstock and Crawley find the same to be true of worker ants of the genus Lasius.

I believe that Phillips, in the remarks above quoted, suggests an important fact which has been too little noticed and may account for much misunderstanding in regard to sex determination in the honey-bee. Accurate knowledge of the life history of a particular individual in a colony of social insects is almost or quite unattainable, for two reasons: first, the egg can not be isolated and the larva brought up by hand, like a young chick, because it requires the presence of nurses of its own species and they will not rear it under abnormal conditions; and second, the workers of many social insects are very fond of eating the eggs and young larvæ and these same workers or the queen not infrequently at once lay eggs in the place of those devoured. This behavior is especially common and disconcerting among the ants. Now a rigid control would require not only that the mother insect should be observed during the very act of oviposition, but that the egg and resulting larva should be kept under constant observation day and night till the completion of development. A relay of observers, changing every few hours for two or three weeks, would therefore be needed in order to make sure that a particular adult had developed from a particular egg, and it would be necessary to observe many individuals in such a manner before we should have the data for accurate conclusions. In fact, we shall need all the resources of a specially equipped laboratory, with a specially trained staff, for any final solution of many of the peculiar developmental problems suggested by the honey-bee and other social insects.

Among these problems we should also have to include that of the differentiation of the two female castes, that is, the problem as to whether the worker and queen arise from one kind or from two different kinds of eggs. The experiments of transferring larvæ of different ages from worker to queen cells, as previously stated, and the existence of series of transitional forms between the worker and the queen, naturally lead to the view that there is only one kind of female egg and that the character of the larval food after the fourth day determines whether the adult is to be a worker or a queen. This may be true of the honey-bee, but, as we shall see, observations on certain ants and termites indicate that there may be more than one kind of female egg.

The main object of the rich and abundant food administered to the larva of the queen honey-bee is evidently the rapid development of her ovaries, so that she may begin to lay eggs very soon after emergence. This is also indicated by the conditions in the Meliponinæ. The Melipona queen, which is reared in a closed cell of the same size as that of the workers and on the same amount of food, emerges with rudimentary ovaries and has to develop them by subsequent feeding during her adult instar, whereas the Trigona queen, which is reared in a large cell with more food than is given to the worker larvæ, emerges with mature or nearly mature eggs in her ovaries. All these queens, however, are distinguished from the cospecific workers by certain degenerate or primitive characters, which, it would seem, must owe their peculiarities either to the indirect, inhibiting or modifying action of chemical substances (enzymes) in their food or to the more direct action of hormones, or internal secretions produced by the developing ovaries. The great size of the ovaries in the queens of all these social bees accounts, of course, for their extraordinary fecundity and the size of their colonies. Cheshire computed the number of eggs which may be laid during her lifetime by a vigorous, fecundated honey-bee queen as about 1,500,000, and, according to von Buttel-Reepen, we should find in her spermatheca no less than 200,000,000 spermatozoa. It is not surprising, therefore, that a hive, at the time of its climax development during the early summer, may contain 50,000 to 60,000 or even 70,000 to 80,000 bees.

In conclusion I may refer to one of the negative peculiarities of social bees — the absence of that peculiar interchange of nutriment between the adult and larva, or trophallaxis, which seems to be a powerful factor in integrating and maintaining the colonies of the social wasps. Among the Meliponinæ the food and egg are simply sealed up in the cell, so that there can be no contact between adults and larva, and even the honey-bee worker does not place the food on the mouth of the larva but pours it on the bottom of the cell where it can be imbibed when needed. So far as known, the bee larva, unlike the wasp larva, produces no salivary secretion to attract its nurses, though it might be going too far at the present time to say that this is certainly not the case. It is quite probable, nevertheless, that the sources of the development and perpetuation of adult and larval contact, so essential to the maintenance of social life among the Bombinæ, Meliponinæ and Apinæ, are to be looked for in other directions. Hermann Müller long ago pointed out, as I stated in the preceding lecture, that the transition of the adult wasp from an insect to a nectar and pollen diet was due to economy of food.[17] These latter substances represent a very concentrated and energizing food supply and one that can be more readily obtained in great abundance than insect food. Hence it is not surprising that a large group of insects like the bees has become so exquisitely anthophilous, and that the exploiting of larval secretions is unnecessary. It will be noticed that all three subfamilies of social bees store quantities of pollen and honey in open cells and such easily accessible stores of liquid and very finely divided food make even the reciprocal feeding of the adult workers in bee colonies superfluous. This storage of food may be at least one of the reasons why such exchanges of nutriment as we observed among the social wasps and shall see again in a more exaggerated form among the ants and termites, were either never developed or were long ago discontinued by the social bees.

ANTS, THEIR DEVELOPMENT, CASTES, NESTING AND FEEDING HABITS

ON one occasion several years ago when I was about to lecture on ants in Brooklyn, a gentleman introduced me to the audience by quoting the sixth to eighth verses of the sixth chapter of Proverbs, and then proceeded in utter seriousness to give an intimate account of their author. He said that Solomon was the greatest biologist the Hebrews had produced, that he had several large and completely equipped laboratories in which he busied himself throughout his reign with intricate researches on ant behavior and that the 700 wives and 300 concubines mentioned in the Bible were really devoted graduate students, who collaborated with the king in his myrmecological investigations. The gentleman deplored the fact that the thousand and one monographs embodying their researches had been lost, and concluded by saying that he was delighted to introduce one who could supply the missing information. As he had consumed just forty-three minutes with his account of Solomon and his collaboratrices, I had to confess my inability to " deliver the goods " in the remaining seventeen. From what recondite sources of biblical exegesis the Brooklyn gentleman drew his information I have never been able to ascertain, but I am sure that Solomon's few myrmecological comments, which have come down to us from about 970 B.C., are very accurate — far more accurate than that story of Herodotus, written some 500 years later, of the gold-digging ants of India, which were as large as leopards, and whose hides were seen by Nearchus in the camp of Alexander the Great, and whose horns were mentioned by Pliny as hanging, even in his time, in the temple of Hercules at Erythræ.[1] This and the many other ant stories invented or disseminated by ancient and modern writers are certainly not devoid of interest, but the actual behavior of the insect is so much more fascinating that you will pardon me for not dwelling on them.

The Formicidæ constitute the culminating group of the stinging Hymenoptera and have attracted many investigators for more than a century and especially during the past thirty years. Unlike the honey-bees these insects make no appeal to our appetites nor even to that vague affection which we feel for most of the common denizens of our forests, fields and gardens, but only to our inquisitiveness and anxiety. Hence the vast literature which has been written on the ants may be said to have been prompted by scientific, philosophic or mere idle curiosity or by our instinct of self-preservation.[2] In the presence of the ant we experience most vividly those peculiar feelings which are aroused also by many other insects, feelings of perplexity and apprehension, which Maeterlinck [3] has endeavored to express in the following words: " The insect does not belong to our world. Other animals and even the plants, despite their mute lives and the great secrets they enfold, seem not to be such total strangers, for we still feel in them, notwithstanding all their peculiarities, a certain terrestrial fraternity. They may astonish or even amaze us at times, but they do not completely upset our calculations. Something in the insects, however, seems to be alien to the habits, morals and psychology of our globe, as if it had come from some other planet, more monstrous, more energetic, more insensate, more atrocious, more infernal than our own. With whatever authority, with whatever fecundity, unequaled here below, the insect seizes on life, we fail to accustom ourselves to the thought that it is an expression of that Nature whose privileged offspring we claim to be. . . . No doubt, in this astonishment and failure to comprehend, we are beset with an indefinable, profound and instinctive uneasiness, inspired by beings so incomparably better armed and endowed than ourselves, concentrations of energy and activity in which we divine our most mysterious foes, the rivals of our last hours and perhaps our successors. . . ."

The similarities which the ants, as one of several families of aculeate, or stinging Hymenoptera, necessarily bear to the wasps and bees, are so overlaid by elaborate specialization, and idiosyncrasies that their primitive vespine characters are not very easily detected. I wish to dwell on some of these specializations, but before doing so, it will be advisable to give under separate captions a brief summary of what I conceive to be the fundamental peculiarities of the ants:

1. The whole family Formicidæ consists of social insects, that is, it includes no solitary nor subsocial forms such as we found among the beetles, wasps and bees. We are therefore unable to point to any existing insects that might represent stages leading up to the social life of the ants. Within the family, nevertheless, we can distinguish quite a number of stages in a gradual evolution of social conditions from very simple, primitive forms, whose colonies consist of only a few dozen individuals, with a comparatively feeble caste development, to highly specialized forms

with huge colonies, comprising hundreds of thousands of individuals and an elaborate differentiation of castes.

2. The number of described species of ants is approximately 3,500, but if we include their subspecies and varieties, many of which will probably be raised to specific rank by future, less conservative generations of entomologists, we shall have more than double that number. This is far in excess of the number of all other social insects, including both the groups I have already considered and the termites. The ants are therefore the dominant social insects.

3. This dominance is shown also by their geographical distribution, which is world-wide. There are ants everywhere on the land-masses of the globe, except in high arctic and antarctic latitudes and on the summits of the higher mountains. The number of individual ants is probably greater than that of all other insects. With few exceptions, the termites are all confined to tropical or subtropical countries, and the number of social wasps and bees in temperate regions is very small.

4. We found that the social wasps arose from the Eumenine solitary wasps and the bees from the solitary Sphecoids. All the authorities agree that the ants had their origin in neither of these ancestral stocks, but among the Scolioids, a distinct offshoot of the primitive Vespoids. Of the four modern families of the Scolioids, the Psammocharidæ, Thynnidæ, Mutillidæ and Scoliidæ, the last seems to be most closely related to the ants. Since they must be traced to ancestors which were winged in both sexes, the Thynnids and Mutillids, which have wingless females, are excluded, and the family Psammocharidæ is not very closely allied to the Formicidæ.

5. The ants, unlike the social wasps and bees, are em-

inently terrestrial insects. They inherited and seem very early to have exaggerated the terrestrial habits of their primitive Scolioid ancestors. The majority of the species in all parts of the world still nest in the soil. Many of them later took to nesting in dead or decaying wood, and more recently a number of species, especially in the rain-forests of the tropics, have become arboreal and nest by preference in the twigs of trees and bushes or construct paper or silken nests among the leaves and branches. The terrestrial habit led to a permanent phylogenetic suppression of the wings in the workers, an ontogenetic loss of the wings in the queens and a diminution of the eyes in both of these castes. A few very archaic ants still possess large eyes like the wasps and bees, but in the great majority of species, which are more or less subterranean, and therefore practically cave-animals during much of their lives, the eyes have dwindled, and in many species have almost or completely disappeared. The great abundance of ants in the desert, savanna and prairie regions of the globe indicates that they arose during some period of the Mesozoic, perhaps during the Triassic or Liassic, when the climate was warm but arid. Their extensive adaptation to low, damp jungles, with their rank vegetation, seems to have developed during the Cretaceous or early Tertiary. The ants therefore resemble the solitary wasps, which are still conspicuously abundant in hot, arid regions. Both groups are represented by only a small number of species in cool, moist regions like New Zealand, the British Isles and certain mountain ranges, like the Selkirks of British America.

6. In the social wasps and bees we found that the worker, or sterile caste, though distinctly differentiated, is, nevertheless, very much like the queen, or fertile female. In ants the differences are much greater. Even when, as in many

primitive ants (Fig. 55), the worker resembles the queen in
size and form, it never possesses wings, and in most ants the
two castes are so dissimiliar that they have often been described as separate species. The male ant, too, is much less
like the queen than is the corresponding sex among the

Fig. 55.—*Stigmatomma pallipes* a primitive, subterranean Ponerine
ant of the United States. The winged individuals are virgin queens and
are very similar to the workers. Nearly twice natural size. (Photograph
by J. G. Hubbard and O. S. Strong.)

social wasps and bees (Fig. 57). It is evident, therefore,
that all three castes are more highly specialized. In many
ants, as we shall see, the worker, queen and male may each
become differentiated into two or more castes, a phenomenon which is nowhere even suggested among the wasps
and bees.

7. Very long and intimate contact with the soil has made

the ants singularly plastic in their nesting habits. While
most social wasps and bees construct elaborate combs with
very regular, hexagonal cells of such expensive substances
as paper and wax, the ants merely make more or less irreg-
ular galleries or chambers in the soil or dead wood or if they
construct paper or silken nests avoid a rigid type of architecture. Hence the great variability of nesting habit in the
same species. This plasticity and saving of time and labor
are very advantageous, because they enable the insects,
when conditions of temperature or moisture become unfavorable or when bothersome enemies settle too near the nest,
to change their habitation readily and without serious loss
to the colony. Espinas long ago noticed the importance of
the terrestrial habits of ants.[4] He says: " Ants owe their superiority to their terrestrial life. This assertion may seem
paradoxical, but consider the exceptional advantages af-
forded by a terrestrial compared with an aërial medium in
the development of their intellectual faculties! In the air
there are the long flights without obstacles, the vertiginous
journeys far from real bodies, the instability, the wandering
about, the endless forgetfulness of things and of oneself.
On the earth, on the contrary, there is not a movement that
is not a contact and does not yield precise information, not a
journey that fails to leave some reminiscence; and as
these journeys are determinate, it is inevitable that a portion of the ground incessantly traversed should be regis-
tered, together with its resources and its dangers, in the
animal's imagination. Thus there results a closer and
much more direct communication with the external world.
To employ matter, moreover, is easier for a terrestrial
than an aërial animal. When it is necessary to build, the
latter must, like the bee, either secrete the substance of its
nest or seek it at a distance, as does the bee when she collects

propolis, or the wasp when she gathers material for her paper.   The terrestrial animal has its building materials close at hand, and its architecture may be as varied as these materials.   Ants, therefore, probably owe their social and industrial superiority to their habitat."

8. The plasticity of ants is shown even more clearly in their care of their young, which are not reared in separate cells but in clusters and lie freely in the chambers and galleries of the nest where they can be moved about and easily carried away or hidden when the colony is disturbed or the moisture and temperature conditions are unfavorable. Like their continual contact with their physical environment, their intimate acquaintance with their young in all their stages has been an important factor in the high psychological development of the Formicidæ.

9. A similar plasticity characterizes their feeding habits. As a group they feed on an extraordinary range of substances: the bodies and secretions of other insects, seeds, delicate fungi, nectar, the saccharine excreta of plant-lice, scale insects, etc.   Some species seem to be almost omnivorous.

10. All this adaptability, or plasticity in nesting and feeding habits is, of course, an expression of a very active and enterprising disposition and has resulted in the formation of a vast and intricate series of relationships between ants and other organisms, including man.   These restless, indefatigable, inquisitive busybodies, forever patrolling the soil and vegetation in search of food, poke their noses, so to speak, into the private affairs of every living thing in their environment.   Nor do they stop at this; they actually draw many organisms, by domesticating them or at any rate attaching them to their nests or bodies, into the vortex of their ceaseless, impudent activities.   Nearly every week

during the past twenty years I have received from some entomologist somewhere on our planet one or more vials of ants with a request for their identification, often because they had been found associated with some insect or plant which the sender happened to be investigating. In the next lecture I shall describe a number of the strange partnerships into which ants have entered as a result of their inordinate and unappeasable appetites.

As my time is limited I shall select for discussion only a few of the topics suggested in the foregoing summary, namely, the main taxonomic divisions of the family Formicidæ, polymorphism, or the development of castes, the origin and growth of colonies, the structure of the alimentary canal in adult and larval ants and the evolution of the feeding habits.

In their main outlines, at least, the phylogenetic relationships of the various subdivisions or subfamilies of the Formicidæ have been clearly established. There are seven of them: the Ponerinæ, Cerapachyinæ, Dorylinæ, Pseudomyrminæ, Myrmicinæ, Dolichoderinæ and Formicinæ.[5] The Ponerinæ constitute the primitive, basic stock of the family and have given rise to the six other subfamilies, which are represented in the ancestral tree (Fig. 56) as so many branches. Their thickness roughly indicates their vigor or comparative development and their height their degree of specialization and dominance in the existing fauna. All the subfamilies are well represented in the tropics of both hemispheres, but in the north temperate region nearly all the species belong to the two largest and highest subfamilies, the Myrmicinæ and Formicinæ. In temperate North America and Eurasia there are very few Dolichoderinæ and Ponerinæ and no Cerapachyinæ nor Pseudomyrminæ. A small number of Dorylinæ extend as far north as

Colorado, Missouri and North Carolina (35° to 40°) and to about the same latitude on the southern shores of the Mediterranean.

With the exception of a series of peculiar parasitic genera, which are represented only by males and females,

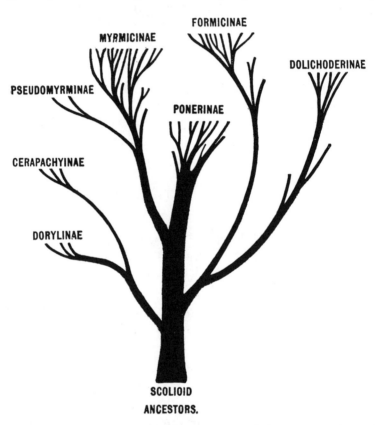

Fig. 56. — Ancestral tree showing the putative phylogenetic relations of the family Formicidæ as a whole and of its subfamilies to one another.

all ants possess a sharply defined worker caste. In primitive groups, like the Ponerinæ (Fig. 55), Cerapachyinæ and Pseudomyrminæ, the worker is nearly as large as the queen

but lacks the wings and has therefore a more simply constructed thorax, the compound eyes are smaller and the
simple eyes, or ocelli, are minute or absent.   In the three
subfamilies mentioned the worker is monomorphic, that

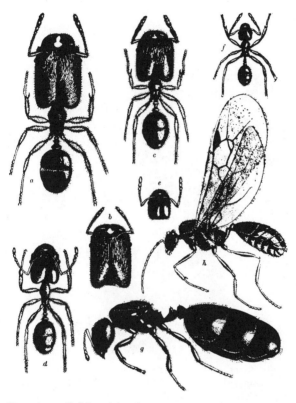

FIG. 57. — A small Myrmicine harvesting ant of Texas, *Pheidole instabilis*, with polymorphic worker caste. *a*, soldier; *f*, worker; *b* to *e*, forms intermediate between the soldier and worker (lacking in most other species of the huge genus Pheidole); *g*, queen (deälated); *h*, male. The figures are all drawn to the same scale.

is, it always has the same form though it may vary somewhat in size.   In the four remaining subfamilies (Dory-

linæ, Myrmicinæ, Dolichoderinæ and Formicinæ) we find
the same uniformity of the worker in many species, but in
a considerable number it has become highly variable, or
polymorphic, as a result of agencies which have acted independently in each subfamily or even within the limits of

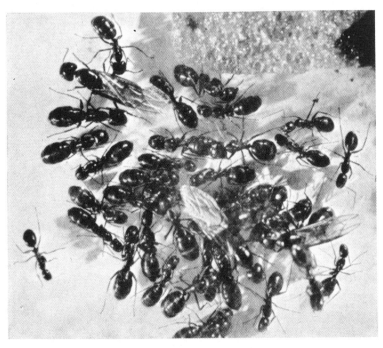

Fig. 58. — Portion of a colony of a common Formicine ant (*Camponotus
americanus*), comprising virgin, winged queens and workers, the latter
showing the unstable polymorphism in stature and size of head, characteristic of most species of the genus. (Photograph by J. G. Hubbard
and O. S. Strong.)

a single genus (Figs. 57 and 58).   In such cases the workers
can be arranged in a graduated series, beginning with
large, huge-headed individuals, more like the queen in
stature, and ending with minute, small-headed individuals,
which may be very much smaller than the queen.   Such a

series exhibits not only great morphological but also great functional differences among its members. The largest individuals commonly act as policemen or defenders of the colony, but in some species their powerful jaws enable them to crush seeds or the hard parts of insects, so that the softer parts may be exposed and eaten by the smaller individuals (Fig. 57). The latter excavate the nest, forage for food, nurse the young and in some species devote all their energies to the cultivation of the fungus gardens. In a graduated series like the one described we usually call the largest workers " maximæ," the smallest " minimæ " and the intermediate forms " mediæ," the word " operariæ " (workers) being understood in each case. Now in some ants only the two extremes, the maximæ and the minimæ, of the polymorphic series proved to be serviceable to the colony, so that all the intermediate forms (mediæ) have been eliminated, leaving the worker caste distinctly dimorphic. In such ants we call the maximæ " soldiers " (*milites*) and the minimæ " workers " (*operariæ*). This condition has been attained in several genera and subgenera among the Myrmicinæ and Formicinæ (Pheidole (Fig. 57), Oligomyrmex, Colobopsis, etc.). In still other genera, where soldiers were not needed or were too expensive to rear and maintain, on account of their great size and appetites, they too have been eliminated and the worker caste is represented only by the tiniest individuals of the originally polymorphic series (Carebara, Tranopelta, Pædalgus, Solenopsis, etc.). There is therefore an enormous difference in these ants in size and structure between the queen and the only surviving worker form of the species. In Carebara, *e.g.*, the queen is several thousand times as large as the worker! Nevertheless, both are merely extreme female forms of the same species and may, of course, develop from the eggs of the same mother. (See Fig. 80, p. 203.)

But the worker is not the only caste that has become dimorphic. In some species there are two distinct forms of queen, in others two distinct forms of male. In these cases one of the forms is winged, the other usually apterous. And here again, by suppression of the winged female, or winged male, the wingless form may become the only surviving fertile form of its sex in the species. All these developments are interesting because they indicate that the distinctions among the various castes have arisen gradually by continuous or fluctuating variations and that the survival and persistence of some of them and the elimination of others have led to the sharply discontinuous series of castes which we find in many ants.   .

It is obvious that some of the differences between the various castes, especially those in size, are due to differences in the amount of food consumed during the larval stages, but the profounder morphological differences which separate the queens, soldiers and workers, must be due to other causes. We must suppose either that the food administered to the larvæ differs in quality or that there are several different kinds of eggs, some of which develop into fertile, others into sterile forms. In a sense the latter would be mutations, like the various sterile forms of the evening primrose, which make their appearance generation after generation from some of the seeds of the fertile forms. In the case of the ants, however, we find that the workers not infrequently lay viable eggs, and though they are never fertilized and generally develop into males, the latter may mate with queens and thus be a means of establishing a representation of the characters of their worker mothers in the germ-plasm of the species. The peculiar anomalies known as gynandromorphs, that is, individuals partly male and partly female, which occasionally occur among ants, also indicate that queens, soldiers and workers arise from as many different kinds of eggs, since there are three different kinds of gynandromorphs, exhibiting respectively combinations or mosaics of male and queen, male and soldier and male and worker characters. It is difficult to see how such perfectly definite combinations could be produced by larval feeding, and it is equally difficult to account for them as the results of internal secretions. In the present state of our knowledge we can only surmise that the differences between the queen and worker castes were originally ontogenetic and determined by feeding, as they still are in the social wasps and bees, but that in the ants the germ-plasm has somehow been reached and modified, so that an heredity basis for caste differentiation has been established.

The ant colony may be initiated and developed by one of two different methods which I shall call the independent and the dependent. The former is peculiar to the non-parasitic, the latter to the parasitic species. Leaving an account of the ants which employ the dependent method for the next lecture, I would say that the great majority of ants establish their colonies in essentially the same manner as Vespa and the bumble-bees. The winged, virgin queen, after fecundation during her nuptial flight, descends to the ground, rids herself of her wings and seeks out some small cavity under a stone or piece of bark, or excavates a small cell in the soil. She then closes the opening of the cell and remains a voluntary prisoner for weeks or even months while the eggs are growing in her ovaries. The loss of the wings has a peculiar effect on the voluminous wing-muscles in her thorax, causing them to break down and dissolve in the blood plasma. Their substance is carried by the circulation to the ovaries and utilized in building up the yolk of the eggs. As soon as the eggs mature, they are laid and the queen nurses the hatching larvæ and feeds them with her saliva till they pupate. Since she never leaves the cell during all this time and has access to no food, except the fat she stored in her abdomen during her larval life and her dissolved wing-muscles, the workers that emerge from the pupæ are all abnormally small. They are, in fact, always minimæ in species which have a polymorphic worker caste. They dig their way out through the soil, thus establishing a communication between the cell and the outside world, collect food for themselves and their mother and thus enable her to lay more eggs. They take charge of the second brood of eggs and larvæ, which, being more abundantly fed, develop into larger workers. The population of the colony now increases rapidly, new chambers and galleries are added to the nest and the queen devotes herself to digesting the food received from the workers and to laying more eggs. In the course of a few years numerous males and queens are reared and on some meteorologically favorable day the fertile forms from all the nests of the same species over a wide expanse of country escape simultaneously into the air and celebrate their marriage flight. This flight provides not only for the mating of the sexes but also for the dissemination of the species, since the daughter queens, on descending to the ground, usually establish their nests at some distance from the parental colony.

It will be seen that the queen ant, like the queen wasp and bumble-bee but unlike the queen honey-bee, is the perfect female of her species, possessing not only great fecundity but in addition all the worker propensities, as shown by her ability to make a nest and bring up her young. But as soon as the first brood of workers appears, these propensities are no longer manifested. That they are not lost is shown by the simple experiment of removing the queen's first brood of workers. Then, provided she be fed or have a sufficient store of food in her body, she will at once proceed to bring up another brood in the same manner as the first, although she would have manifested no such behavior under normal conditions.

As already stated, this independent method of colony formation is the most universal and is followed alike by tropical and extra-tropical ants. It is undoubtedly the primitive method and, as we shall see, the one from which the dependent method has been derived. It differs from that of Vespa and Bombus, nevertheless, in leading to the formation of perennial colonies even in temperate and boreal regions. The queen ant may, in fact, live from 12 to 17 years and although, like other aculeates, she is fecundated only once, may produce offspring up to the time of her death. Unlike the queen honey-bee she is never hostile to her own queen daughters, and in many species of ants some of these daughters may return after their marriage flight to the maternal colony and take a very active part in increasing its population. In this manner the colony may become polygynic or pleometrotic, and in some instances may contain a large number of fertile queens. When such a colony grows too large it may separate into several, the queens emigrating singly or in small companies, each accompanied by a detachment of workers, to form a new nest near the parental formicary. This behavior is exhibited by the well-known mound-building ant (*Formica exsectoides*) of our New England hills. You will notice that its mounds usually occur in loose groups or clusters and that the workers of the different nests are on friendly terms with one another and sometimes visit back and forth. We may, of course, call the whole cluster a single (poly-domous) colony, but it really differs from a number of colonies only in the absence of hostility between the inhab-itants of the different mounds. In certain tropical ants, like the Dorylinæ (Figs. 59 and 60), however, I am inclined

FIG. 59. — Argentinian legionary (Doryline) ant *Eciton* (*Acamatus*) *strobeli*. Workers showing polymorphism, and male, photographed to the same scale as the four smaller workers. (Photograph by Dr. Carlos Bruch.)

to believe that the only method of colony formation is by a splitting of the original colony into as many parts as it contains young queens. These huge, clumsy creatures (Fig. 60) are always wingless and must therefore be fecun-dated in the nest, and since the colonies, which comprise hundreds of thousands of workers, are nomadic and keep wandering from place to place, they must become inde-pendent entities as soon as they are formed.

We possess no accurate data on the age that ant colonies may attain. Some of them certainly persist for 30 or 40 years and probably even longer. In such old colonies the original queen has, of course, been replaced by successive generations of queens, that is, by her fertile daughters, granddaughters and great-granddaughters, and the worker

Fig. 60. — Dorsal and lateral view of the wingless queen (dichtha-diigyne) of *Eciton* (*Acamatus*) *strobeli*. Same scale as Fig. 59. (Photograph by Dr. Carlos Bruch.)

personnel has been replaced at a more rapid rate, because the individual worker does not live more, and in most instances lives considerably less, than three or four years.

The feeding habits of ants are so varied and complicated that it will be advisable before considering them to describe the structure of the alimentary canal in both adult and larva.[6] The mouth-parts of the adult are of the generalized vespine type and consist of a small, flap-like upper lip, or labrum, a pair of strong, usually toothed mandibles, a pair of small maxillæ and a broad lower lip, or labium. The

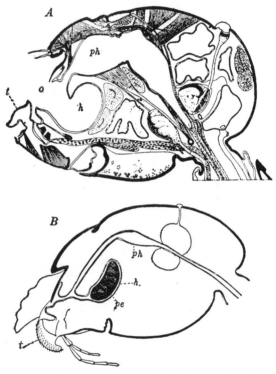

FIG. 61. — Sagittal sections through the heads of ants. *A*, of queen *Lasius niger* with the mouth open (After Janet). *B*, of queen *Camponotus brutus* with the mouth closed. *t*, tongue; *o*, oral orifice; *ph*, pharynx; *h*, infrabuccal pocket; *pe*, pellet *in situ*, made up of solid particles of food refuse and strigil sweepings. Note stratification in the substance of the pellet, indicating successive meals or toilet operations.

maxillæ and labium are each provided with a pair of jointed, sensory appendages, the palpi. The mandibles, which are really the ant's hands, vary greatly in shape in

different genera and are used not only in securing the food
but also in many other activities, such as digging in the
earth or wood, transporting other ants or the young, fighting, leaping, etc. Liquids are, of course, merely imbibed
and swallowed, but solid food is seized and crushed with the
mandibles and the juices or smaller particles licked up with
the tongue, which is a roughened pad at the tip of the
lower lip (Fig. 61*t*) just anterior to the opening of the
duct of the salivary glands. The small particles thus collected are carried back into a small chamber or sac, the infrabuccal pocket (Fig. 61*h*), which lies immediately below
and anteriorly to the mouth-opening (*o*). This pocket is
an important structure since it serves as a receptacle not
only for the more solid particles of food but also for the
dirt, fungus-spores, etc., which the ant collects during her
toilet operations, for the ant is an exquisitely cleanly insect
and devotes much of her leisure to licking and burnishing
her own smooth or finely chiseled armor and that of her
nest-mates. Moreover, the tip of the fore tibia is furnished
with a beautiful comb or strigil which can be opposed to
another comb on the concave inner surface of the fore meta-
tarsus. The ant cleans her legs and antennæ by drawing
them between these combs, which are then drawn across
the mouth, with the result that any adhering dirt is carried
off into the infrabuccal pocket. In this manner the dirt
and the solid or semisolid food particles are combined and
the whole mass moulded in the infrabuccal pocket into the
form of a roundish oblong pellet (Fig. 61 *B pe*). After any
liquid which it may contain has been dissolved out and
sucked back into the mouth, the pellet is cast out, so that
no solid food actually enters the alimentary canal. All
adult ants therefore subsist entirely on liquids.

The alimentary canal proper is a long tube extending

through the body and divided into sections, each with its special function. The more anterior sections are the mouth cavity, the pharynx (Fig. 61 *ph*), which receives the ducts of certain glands, and the very long, slender gullet, which traverses the posterior part of the head, the whole thorax and the narrow waist, or pedicel of the abdomen as far as the base of its large, swollen portion, the gaster. Here the gullet expands into a thin-walled, distensible sac, the crop, which is used for the storage of the imbibed liquids. At its posterior end the crop is separated from the ellipsoidal stomach by a peculiar valvular constriction, the proventriculus. The hindermost sections of the alimentary tract are the intestine and the large, pear-shaped rectum. The crop, proventriculus and stomach are the most interesting of these various organs. Forel calls the crop the " social stomach," because its liquid contents are in great part distributed by regurgitation to the other members of the colony and because only a small portion, which is permitted to pass back through the proventricular valve and enter the stomach, is absorbed and utilized by the individual ant. That the crop functions in the manner described can be readily demonstrated by permitting some pale yellow worker ant to gorge herself with syrup stained blue or red with an aniline dye. The ant's gaster will gradually become vividly colored as the crop expands. Now if the insect be allowed to return to the nest, other workers will come up to it, beg for food with rapidly vibrating antennæ and protrude their tongues, and very soon their crops, too, will become visible through the translucent gastric integument as they fill with the stained syrup. Then these workers in turn will distribute the food by regurgitation in the same manner till every member of the colony has at least a minute share of the blue or red cropful of the first worker.

The alimentary tract of the helpless, legless, soft-bodied ant grub or larva is much simpler than that of the adult. The mouth-parts are similar but more rudimentary. As a rule, the mandibles are less developed but in some larvæ they are strong, dentate and very sharp. The lower lip is fleshy and protrusible and provided with sensory papillæ instead of palpi, and the unpaired duct of the long, tubular and more or less branched salivary glands opens near its tip. The mouth-opening is broad and its lining in many species is provided with numerous transverse ridges beset with very minute spinules (Fig. 62 C). Larger, pointed projections or imbrications may also cover the basal portions of the mandibles. All these spinules and projections are probably used in triturating the food but perhaps when rubbed on one another they may also produce shrill sounds for the purpose of apprising the worker nurses of the hunger or discomfort of their charges. The gullet is long and very slender and opens directly into the large stomach, which throughout larval life is closed behind, that is, does not open into the intestine. A communication with the more posterior portion of the alimentary tract is not established till the larva is about to pupate. Then all the undigested food which has accumulated in the stomach since the very beginning of larval life is voided as a large black pellet, the meconium.

In the larvæ of the Pseudomyrminæ (Figs. 62, 63 and 64) there are certain very peculiar additional structures which may be briefly described. The head is not at the anterior end of the body as in other ant larvæ but pushed far back on the ventral surface so that it is surrounded by a great hood formed from the three thoracic segments, and the first abdominal segment, which lies immediately behind the head, has in the midventral line a singular pocket, the

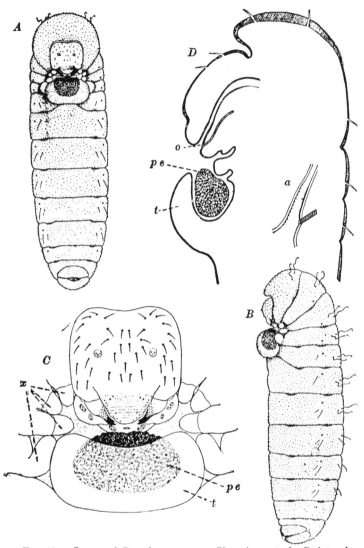

FIG. 62. — Larva of *Pseudomyrma gracilis*. *A*, ventral; *B*, lateral view; *C*, head and adjacent portions of same enlarged; *D*, sagittal section through anterior portion of larva. *o*, oral orifice; *x*, exudatoria; *t*, trophothylax, or pocket, which holds the pellet (*pe*), deposited by the worker nurses and which is eaten by the larva. Note the hooked dorsal hairs of the larva, which serve to suspend it from the walls of the nest. *a*, mouth cavity, more enlarged to show the fine spinules (also seen in *C*), which serve to triturate the pellet and probably also as a stridulatory organ.

trophothylax (t). Furthermore, each side of this segment and each ventrolateral portion of the several thoracic segments is developed as a peculiar protuberance or appendage, which functions as a blood-gland, or exudatorium (x).

Unlike the adult ants the larvæ can devour solid food, though they are often fed, at least in their youngest stages,

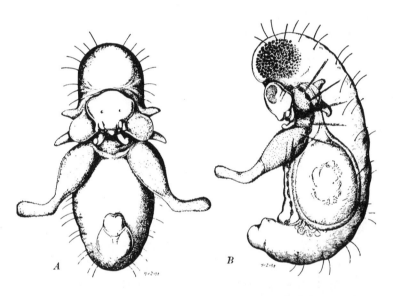

FIG. 63.—A, ventral; B, lateral view of the first larval stage ("trophidium") of the Ethiopian *Pachysima latifrons,* showing the peculiar appendages ("exudatoria") surrounding the head. These belong to the three thoracic and the first abdomonial segments.

with liquids regurgitated on their mouths by the worker nurses. The larvæ of the Pseudomyrminæ are fed with the pellets (pe) from the infrabuccal pocket, which are placed by the workers in the trophothylax where they are within easy reach of the mandibles and can be gradually drawn into the mouth, triturated and swallowed. Some primitive ants (Ponerinæ, some Myrmicinæ, etc.) actually feed their young with pieces of insects or entire small insects, which are simply placed on the ventral surface of the larva within reach of its mouth-parts.

In a former lecture I referred to the fact that the larvæ of the social wasps, either before or after feeding, produce droplets of a sweet salivary secretion, which are eagerly imbibed by the adult wasps, and I designated this interchange of food between adult and larva as trophallaxis. I have recently made some observations which show that the ant larvæ also produce secretions which appeal to the appetites of their nurses. These secretions are more varied than in the wasps. Certain ant larvæ undoubtedly supply their nurses with saliva, but many or all sweat a fatty secretion through the delicate general integument of the body, and the larval Pseudomyrminæ produce similar exudates from the papillæ or appendages above described. Although these various substances are produced in very small quantities they are of such qualities that they are eagerly sought by the adult ants. This explains much of the behavior which has been attributed to maternal affection on the part of the queen for the workers, such as the continual licking and fondling of the larvæ, the ferocity with which they are defended and the solicitude with which they are removed when the nest is disturbed. In other words, a decidedly egoistic appetite, and not a purely altruistic maternal anxiety for the welfare of the young constitutes the potent "drive" that initiates and sustains the intimate relations of the adult ants to the larvæ, just as the mutual regurgitation of food initiates and sustains similar relations among the adult workers themselves.

I am convinced that trophallaxis will prove to be the key to an understanding not only of the behavior I have briefly outlined but also of the relations which ants have acquired to many kinds of alien organisms. In the accompanying diagram (Fig. 65) I have endeavored to indicate how trophallaxis, originally developed as a mutual trophic rela-

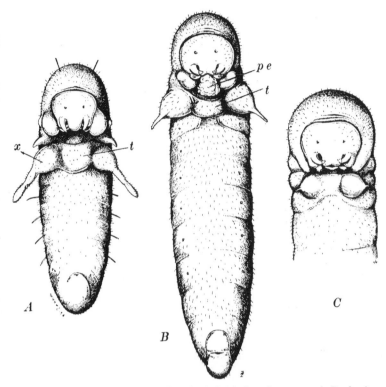

Fig. 64. — Second, third, and fourth (*adult*) larval stages of *Pachysima latifrons,* showing the gradual dwindling of the exudatoria. *A* and *B* show the trophothylax (*t*); and *B* also shows the food pellet *pe,* which is the pellet formed in the infrabuccal pocket of the worker nurse; *x,* exudatorium. See Figs. 62 and 63.

tion between the queen ant and her brood, has expanded with the growth of the colony, like an ever-widening vortex, till it involves, first, all the adults as well as the brood and therefore the entire colony; second, a great number of alien insects that have managed to get a foothold in the

nest as scavengers, predators and parasites (symphiles);
third, alien social insects, that is, other species of ants
(social parasites); fourth, alien insects that live outside the
nest and are " milked " by the ants (trophobionts), and
fifth, certain plants that are regularly visited or even in-

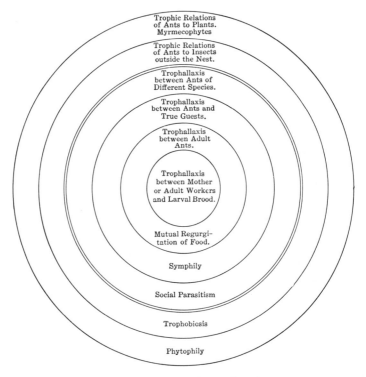

Fig. 65. — See text for explanation.

habited by the ants (myrmecophytes). These extranidal
relationships, represented by the two outer rings in the
diagram are, of course, incomplete or one-sided, since the
organisms which they represent are not fed but merely
cared for or protected by the ants. In my next lecture I
shall have more to say about some of these relationships.

There is throughout the animal kingdom, as I believe Espinas was the first to remark, a clear correlation on the one hand between a solitary life and carnivorous habits and on the other hand between social habits and a vegetable diet. The beasts and birds of prey, the serpents, sharks, spiders and the legions of predacious insects mostly lead solitary lives, whereas the herbivores, rodents, granivorous and frugivorous birds and plant-eating snails and insects are more or less gregarious. Man himself is quite unable to develop populous societies without becoming increasingly vegetarian. Compare, for example, the sparse communities of the carnivorous Esquimaux with the teeming populations of the purely vegetarian Hindoos. The reasons for these correlations are obvious, for plants furnish the only abundant and easily obtainable foods and, at least in the form of seeds and wood, the only foods that are sufficiently stable to permit of long storage. In the previous lectures I have shown that the social beetles and bees are strictly vegetarian and that the social wasps, though descended from highly predatory ancestors, are nevertheless becoming increasingly vegetarian like the bees. The ants exhibit in the most striking manner the struggle between a very conservative tendency to retain the precarious insectivorous habits of their vespine ancestors and a progressive tendency to resort more and more to a purely vegetable regimen as the only means of developing and maintaining populous and efficient colonies. Anthropologists have distinguished in the historical development of human societies six successive stages, designated as the hunting, pastoral, agricultural, commercial, industrial and intellectual. Evidently the first three, the hunting, pastoral and agricultural, are determined by the nature of the food and represent an advance from a primitive, mainly flesh-consuming to a largely vegetarian regimen. Lubbock showed that the same three stages occur in the same sequence in the phylogenetic history of the ants. At the present time we are able to give even greater precision to his outlines of this evolution.

All the primitive ants are decidedly carnivorous, that is predatory hunters of other insects. That this must have been the character of the whole family during a very long period of its history is indicated by the retention of the insectivorous habit, in a more or less mitigated form, even in many of the higher ants. Always striving to rear as many young as possible, always hungry and exploring, the ants early adapted themselves to every part of their environment. They came, in fact, to acquire two environments, each peopled by a sufficient number of insects, arachnids, myriopods, etc., to furnish a precarious food-supply. Most of the ants learned to forage on the exposed surface of the soil and vegetation and became what we call epigæic, or surface forms, while a smaller number took to hunting their prey beneath the surface of the soil and thus became hypogæic, or subterranean. Many of the latter are very primitive but their number has been repeatedly recruited from higher genera, which by carrying on all their activities within the soil have found a refuge and surcease from a too strenuous competition with the epigæic species. We have here some very interesting cases of convergence, or parallel development, since the underground habit has caused the workers, which rarely or never leave their burrows, to lose their deep pigmentation and become yellow or light brown and to become nearly or quite blind. As will be evident in the course of my discussion, the tendency towards vegetarianism is apparent among both the epigæic and hypogæic forms.

The ants belonging to the oldest and most primitive sub-families, the Ponerinæ, Dorylinæ and Cerepachyinæ and also to many of the lower genera of Myrmicinæ, feed exclusively on insects and therefore represent the hunting stage of human society. Owing to the difficulty of securing large quantities of the kind of food to which they are addicted, many of the species form small, depauperate colonies, consisting of a limited number of monomorphic workers. Many of these species lead a timid, subterranean life. In the size of their colonies, which may comprise hundreds of thousands of individuals, the Dorylinæ alone constitute a striking exception, but one which proves the rule. These insects, known as driver, army or legionary ants and very largely confined to Equatorial Africa and tropical America, are strictly carnivorous, but being nomadic and therefore foraging over an extensive territory, are able to obtain the amount of insect food necessary to the growth and maintenance of a huge and polymorphic population.[7] They are the famous ants whose intrepid armies often overrun houses in the tropics, clear out all the vermin and compel the human inhabitants to leave the premises for a time. In Africa they have been known to kill even large domestic animals when they were tethered or penned up and thus prevented from escaping.

The pastoral stage is represented by a great number of Myrmicine and especially of Formicine and Dolichoderine ants which live very largely on " honey-dew." This sweet liquid, concerning the origin of which there was much speculation among the ancients, is now known to be the sap of plants and to become accessible to the ants in two ways. First, it may be excreted by the plants from small glands or nectaries (" extrafloral nectaries ") situated on their leaves or stems, where it is eagerly sought and im-

bibed by the ants. Second, a much more abundant supply
is made accessible by a great group of insects, the Phytophthora, comprising the plant-lice, scale-insects, mealy-bugs,
leaf-hoppers, psyllids, etc., which live gregariously on the
surfaces of plants. These Phytophthora pierce the integument of the plants with their slender, pointed mouth-
parts and imbibe the juices, which consist of water containing in solution cane sugar, invert sugar, dextrin and
a small amount of albuminous substance. In the alimentary canal of the insects much of the cane sugar is
split up to form invert sugar and a relatively small amount
of all the substances is assimilated, so that the excrement
is not only abundant but contains more invert and less
cane sugar. This excrement, or honey-dew either falls upon
the leaves and is licked up by the ants or is imbibed by
them directly while it is leaving the bodies of the Phytophthora. Many species of ants have learned how to induce
the Phytophthora to void the honey-dew by stroking them
with the antennæ, to protect and care for them and even to
keep them in specially constructed shelters or barns. Some
ants have acquired such vested interests in certain plant-
lice that they actually collect their eggs in the fall, keep
them in the nests over winter and in the spring distribute
the hatching young over the surface of the plants. Lin-
næus was therefore justified in calling the plant-lice the
dairy-cattle of the ants (" *hæ formicarum vaccæ* "). This
dairy business is, in fact, carried on in all parts of the world
on such a scale and with so many species of Phytophthora
that it constitutes one of the most harmful of the multi-
farious activities of ants. Their irrepressible habit of protecting and distributing plant-lice, scale-insects, etc., is a
source of considerable damage to many of our cultivated
plants and especially to our fruit-trees, field and garden

crops. Ants mostly attend Phytophthora on the leaves and shoots of plants, but quite a number of species are hypogæic and devote themselves to pasturing their cattle on the roots. Thus our common garden ant (*Lasius americanus*) distributes plant-lice over the roots of Indian corn.

The habit of keeping Phytophthora was probably developed independently in many different genera, and it is easy to see how the habit of feeding by mutual regurgitation among the ants themselves might have led to the behavior I have been describing. Certainly the genera that have developed trophallaxis among the adult members of their colonies are the very ones which most assiduously attend the Phytophthora. And it is equally certain that the latter habit is very ancient, because it was already established among the ants of the Baltic Amber during Lower Oligocene times and that, as we have seen, was many million years ago.

The dairying habit has led to an interesting specialization in certain species known as "honey-ants," which inhabit desert regions or those with long, dry summers.[8] These ants have found it very advantageous to store the honey-dew collected during periods of active plant growth, and as they are unable to make cells like those of wasps and bees, have hit upon the ingenious device of using the crops of certain workers or soldiers for the purpose. In all ants, as we have seen, the crop is a capacious sac, but in the typical honey-ants it becomes capable of such extraordinary distention that the abdomen of the individuals that assume the rôle of animated demijohns or carboys, becomes enormously enlarged and perfectly spherical. Such "repletes" (Fig. 66) are quite unable to walk and therefore suspend themselves by their claws from the ceilings of the nest chambers. When hungry the ordinary workers stroke their heads and receive by regurgitation droplets of the honey-dew with which they were filled during seasons of plenty. The condition here described, or one of less gastric disten-tion, has been observed in desert or xerothermal ants in very widely separated regions and belonging to some nine different genera of Myrmicinæ, Formicinæ and Dolichoder-inæ (Myrmecocystus and Prenolepis in the United States and Northern Mexico, Melophorus, Camponotus, Lepto-myrmex and Oligomyrmex in Australia, Plagiolepis and

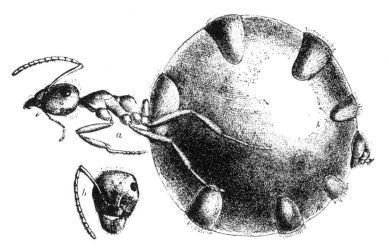

Fig. 66. — Replete of honey-ant (*Myrmecocystus melliger*) from Mexico. *a*, lateral aspect of insect; *b*, head from above.

Aëromyrma in Africa and Pheidole in Australia and the southwestern United States).

A more direct vegetarian adaptation is seen in many Formicidæ that inhabit the same desert or xerothermal regions as the honey-ants. In such regions insect food is at no time abundant and is often so scarce that the ants are compelled to eat the seeds of the sparse herbaceous vegetation. At least a dozen genera, all Myrmicinæ, illustrate this adaptation: Pogonomyrmex, Veromessor, Novomessor and Solenopsis in America, Messor, Oxyopomyrmex, Goniomma, Tetramorium and Monomorium in the southern Palearctic region, Meranoplus in the Indoaustralian, Cratomyrmex and Ocymyrmex in the Ethiopian region and Pheidole (Fig. 57) in the warmer parts of both hemispheres.[9] It was at one time believed that some of these ants actually sow around their nests the grasses and other herbaceous plants from which they gather the seeds, but this has been disproved. They are merely collected, husked and stored in special chambers or granaries in the more superficial and dryer parts of the formicary. Emery has shown that as food the proteins are preferred to the starchy portions of the seeds and are also fed to the larvæ. *Messor barbarus,* the ant to which Solomon refers, is one of these harvesters. Probably none of them disdains insect food when it can be had. Nevertheless the adaptation to crushing hard seeds is so pronounced in certain genera that the mandibles have become distinctly modified. Their blades have become broader and more convex and the head has been enlarged to accommodate the more powerful mandibular muscles. In certain forms (Pheidole, Messor, Novomessor, Holcomyrmex) the soldiers or major workers seem to function as the official seed-crushers of the colony.

The harvesting ants can hardly be regarded as true agriculturists because they neither sow nor cultivate the plants from which they obtain the seeds. Yet there is a group of ants which may properly be described as horticultural, namely the Attiini, a Myrmicine tribe comprising about 100 exclusively American species and ranging from Long Island, N. Y., to Argentina, though well represented by species only within the tropics.[10] The tribe includes several genera (Cyphomyrmex, Apterostigma, Sericomyrmex,

Myrmicocrypta, etc.) the species of which are small and timid and form small colonies with monomorphic workers, while others (Atta and Acromyrmex) are large and aggressive and form very populous colonies with extremely polymorphic workers. The Attas, or parasol-ants inhabit the savannas and forests of South and Central America, Mexico, Cuba and Texas. Their extensive excavations result in the formation of large mounds and often cover a considerable area (Fig. 67). According to Branner, a single mound of the common Brazilian *Atta sexdens* may contain as much as 265 cubic meters of earth, and the population of a colony of this species, according to Sampaio, may number from 175,000 to 600,000 individuals. Of course, the size of the mounds varies with the depth of the excavations, which are much shallower in the rain-forests than in the dry savannas. From their mounds the ants make well-worn paths through the surrounding vegetation and frequently defoliate bushes or trees, cutting large pieces out of their leaves and carrying them like banners to their nests. The pieces are then cut into smaller fragments and built up on the floors of the large nest chambers (Figs. 68 and 69) in the form of sponge-like masses, which become covered with a white, mould-like fungus mycelium (Figs. 70 and 71). The latter is treated in some unknown manner by the smallest, exclusively hypogæic caste of workers, so that the hyphæ produce abundant clusters of small, spherical swellings, the bromatia (Fig. 72), which are eaten by the ants and fed to their larvæ. Each species of Attiine ant cultivates its own particular fungus and no other is permitted to grow in the nest. That the bromatia are really anomalous growths induced by the ants is indicated by the fact that they do not appear when the fungus is grown in isolation on artificial media. Alfred Moeller, who

Fig. 67. — Nest of the Texan leaf-cutting ant (*Atta texana*) at Victoria, Texas. (Photograph by S. J. Hunter.)

Fig. 68. — Vertical section through the center of a nest of the Argentinian leaf cutter, *Atta vollenweideri*, showing the chambers containing the fungus gardens. (Photograph by Dr. Carlos Bruch.)

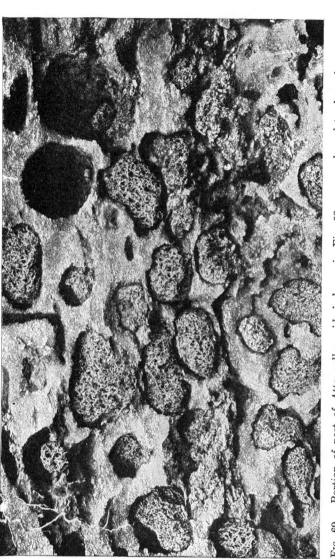

Fig. 69. — Portion of nest of *Atta vollenweideri* shown in Fig. 68, more enlarged to show the sponge-like fungus-gardens *in situ* in the chambers. About one eighth natural size. (Photograph by Dr. Carlos Bruch.)

was the first to cultivate these fungi, regarded them as belonging to the Agarics and named one of them *Rozites gongylophora.*[11] Either the ants prevent the mushrooms from appearing, or, more probably the subterranean conditions under which the mycelium is cultivated are unfavorable to their development. Moeller was also unable to obtain the mushrooms in his cultures, but found those

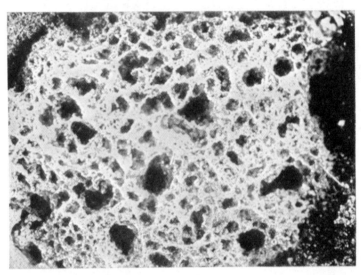

FIG. 70.—Portion of fungus garden of the Texan leaf-cutting ant (*Atta texana*). About one half natural size.

of Rozites growing on the surface of an abandoned Acromyrmex nest. That the fungi cultivated by the various Attiini belong to several different genera is shown by Bruch and Spegazzini who have recently been able to identify the mushrooms of the fungi cultivated by several Argentinian Attiini.[12] *Acromyrmex lundi,* e.g., cultivates *Xylaria micrura* Speg., *Mœllerius heyeri, Poroniopsis bruchi* Speg. and *Atta vollenweideri,* a gigantic Agaric, *Locellina mazzuchii* Speg. (Fig. 73).

The lower genera of the Attiini differ in many particulars from such highly specialized forms as Atta and Acromyrmex. Their nests are smaller and there are differences in the gardens and the substratum, or substances on which the fungi are grown. The species of Trachymyrmex suspend the garden from the ceiling of the nest chamber in-

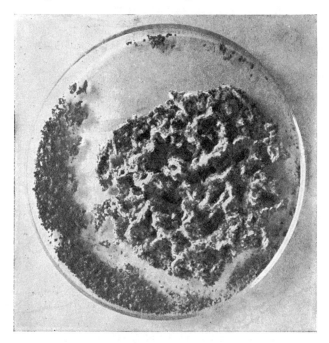

Fig. 71.—Fungus garden built in a Petri dish by a colony of Apterostigma in British Guiana. Natural size. (Photograph by Mr. J. Tee-Van.)

stead of building it on the floor, and in some species of Apterostigma it is enclosed in a spherical envelope of dense mycelium, so that, except for its larger size, it much resembles the silken egg-case of a spider. These ants and others, such as Cyphomyrmex and Myrmicocrypta, use the excrement of other insects, especially of caterpillars, as a substratum for the gardens, and one species, *Cyphomyrmex rimosus,* cultivates a very peculiar fungus (*Tyridiomyces formicarum* Wheeler), which does not grow in the form of a mycelium but of isolated, compact bodies, resembling little pieces of American cheese, and consisting of yeast-like cells.[13]   The same or a very similar genus of fungi is grown by the species of Mycocepurus.

How do all these Attiinie ants come into possession of the various fungi which they cultivate with such consummate

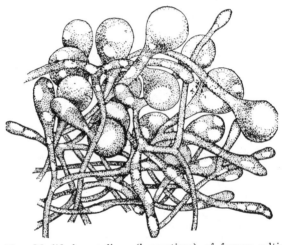

Fig. 72. — Modified mycelium (bromatium) of fungus cultivated by the Argentinian *Moellerius heyeri.* The globular swellings of the hyphæ are produced by the ants.   (After Carlos Bruch.)

skill?   The question is, of course, twofold, since we should like to know how the individual colony obtains its fungus and how the ancestors of the existing Attiini first acquired the fungus-growing habit.   The former question has been answered by the very interesting investigations of Sampaio, H. von Ihering, J. Huber and Goeldi on the Brazilian *Atta sexdens* and of Bruch on the Argentinian *Acromyrmex lundi.*[14]   The virgin queen of these species, before leaving the parental nest for her marriage flight, takes a good meal of fungus. The hyphæ, together with the strigil sweepings from her own body and, according to Bruch, also some particles of the substratum, are packed into her infrabuccal pocket, where they form a large pellet, which she retains till she has mated, thrown off her wings and made a small chamber for herself in the soil. She then casts the pellet

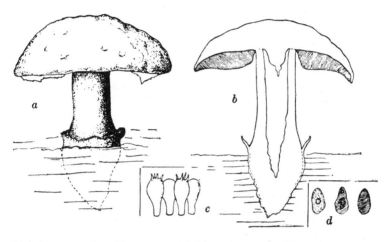

Fig. 73.— a, *Locellina mazzuchii*, the gigantic fruiting phase (pileus 30 to 42 cm. in diameter!) of the fungus cultivated by the Argentinian leaf-cutting ant (*Atta vollenweideri*); b, section of same; c, basidia; d, spores. (After C. Spegazzini.)

on the floor of the chamber where its hyphæ begin to proliferate in the moist air and draw their nutriment from the extraneous materials with which they are mingled (Fig. 74 A). The queen carefully watches the incipient garden and accelerates its growth by manuring it with her feces (C and D). She begins to lay eggs (Fig. 76 A) and even breaks up some of them and adds them to the garden, which soon becomes large enough to form a kind of nest for the intact and developing eggs (Fig. 74 B to F). The young

larvæ on hatching proceed to eat the mycelium and eventually pupate and emerge as small workers, which break
through the soil, bring in pieces of leaves and add them to
the garden. The care of the latter then devolves on the

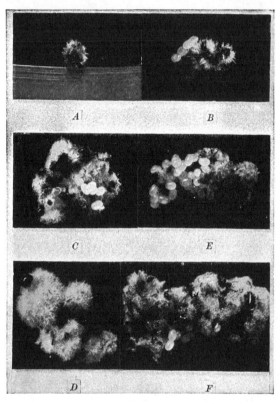

FIG. 74. — Stages in the development of the fungus garden by the
queen of the Argentinian *Mœllerius heyeri.* *A*, pellet of substratum 36 hours after its ejection from the queen's infrabuccal
pocket. The hyphæ have begun to grow. *B*, same pellet after 3
days, with 4 eggs; *C*, same pellet after 8 days, showing droplets of
feces with which the queen manures the hyphæ; *D*, same pellet
after 12 days, also showing droplets of feces; *E*, small fungus
garden after 30 days, with 32 eggs; *F*, same after 40 days. The
magnification of all the figures is very nearly 10 diameters.
(Photographs by Dr. Carlos Bruch.)

workers and the queen henceforth devotes herself to laying
eggs. The colony is now established and its further development is merely a matter of enlarging the nest, multiplying the gardens and increasing the population.   Thus Atta

FIG. 75. — *A*, an infrabuccal pellet of the queen *Mœllerius heyeri*
after cultivation for 36 hours on gelatine. x10. *B*, eggs and pellets
made of filter paper by a queen *Mœllerius heyeri* that had
failed to develop a fungus garden. x10.   (Photograph by
Dr. Carlos Bruch.)

and Acromyrmex transmit their food-plants from generation to generation in a very simple manner, that is, merely

by the queen's retaining, till she has established her nest chamber, the infrabuccal pellet consisting of her last meal in the colony in which she was reared. And there is every reason to suppose that the same method of transmitting the fungus from the maternal to the daughter colonies is practiced by all the other genera of the tribe.

Of course, the answer to the question as to how the ancestors of the Attiini acquired their food-fungi in the first place must be purely conjectural. Yet certain observations by Professor I. W. Bailey [15] and myself seem to indicate from what simple beginnings the elaborate fungus-growing habits may have been evolved. An examination of the infrabuccal pellets of the most diverse ants shows that in nearly every case they contain fungus spores or pieces of mycelium collected from the surfaces of their bodies or from the walls of the nest. Moreover, many ants have a habit of casting their pellets on the refuse heaps, or kitchen-middens of their nests, and Professor Bailey finds that in the case of certain African Crematogasters that live in the moist cavities of plants (Plectronia, Cuviera) the refuse heaps consist very largely of such ejected pellets and produce a luxuriant growth of aërial hyphæ which are cropped by the ants. From such a condition it is, perhaps, only a short step to the establishment of small gardens consisting at first of the pellets and later of these and accumulations of extraneous materials, such as the feces of the ants, those of caterpillars and beetles, vegetable detritus, etc., which might serve to enlarge the substratum and increase the growth of the fungus. The selection of particular species of fungi and their careful culture and transmission are evi-dently specializations that must have been established before the stages represented by even the most primitive ex-isting Attiini could have been attained.

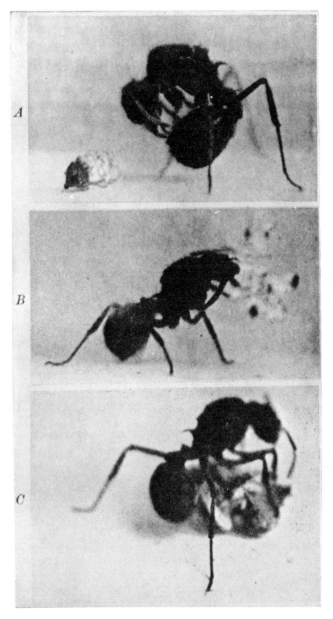

Fig. 76. — Behavior of the queen of *Mœllerius heyeri*. *A*, photographed in the act of laying an egg. The incipient fungus garden in which the egg will be placed is shown to the left resting on the floor of the nest chamber. *B*, queen placing an egg in the fungus garden which is sticking to the glass wall of the artificial nest; *C*, queen photographed in the act of placing a droplet of feces in the fungus garden. Magnification 5 diameters. (Photographs by Dr. Carlos Bruch.)

Whatever may have been the processes whereby the ancestral Attiini developed the fungus-growing habit, it must have originated in the more humid portions of the tropics, since nearly all the more primitive species of the tribe are still confined to the rain-forests. But certain species soon found that by sinking their galleries and

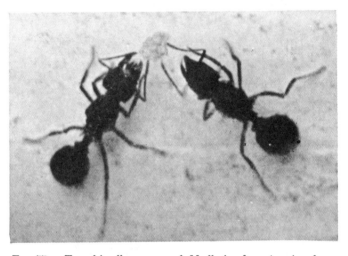

Fig. 77. — Two friendly queens of *Mœllerius heyeri* caring for a single incipient fungus garden, which is adhering to the glass wall of the artificial nest x5. (Photograph by Dr. Carlos Bruch.)

chambers to a greater depth in the soil they could easily carry on their fungus farming even in arid regions. Thus some species of Mœllerius, Trachymyrmex and Cyphomyrmex have come to live in the dry deserts of Arizona, New Mexico and northern Mexico, and as they can always find in such localities enough vegetable material for the substrata of their gardens, they have attained to a control of their environment and food-supply, which even the human inhabitants of those regions might envy.

LECTURE V

PARASITIC ANTS AND ANT GUESTS

THE ants are so favorable for the study of certain phenomena which I have been unable to more than touch on in the
preceding lectures, that I have set this lecture apart for
their fuller consideration. I allude to the phenomena
which biologists embrace under the terms " symbiosis," or
" mutualism " and " parasitism." Social life may, indeed,
be regarded merely as a special form of symbiosis. This
term, which signifies the living together of organisms in a
balanced, coöperative, reciprocally helpful manner, is commonly applied interspecifically, that is, to partners thus
related but belonging to different species, but there is no
reason why it should not be applied to the same kind of
relations between individuals of the same species, that is
intraspecifically. Symbiosis is probably never realized in
its ideal form, which would require that each of the partner
organisms should render to the other in food or services an
exact equivalent of what it receives. So great is the greed
of organisms that one member of the partnership usually
tends to snatch more than its share of the profits accruing
from the association. One member is therefore exploited,
while the other becomes correspondingly dependent, that
is, parasitic. In some groups of animals symbiotic or mutualistic relations may thus lapse into parasitism, but it
seems to me improbable that parasitism among insects has
had such an origin. The common and perhaps exclusive
source of the phenomenon among these highly specialized

organisms is predatism. In fact, the most typical of parasitic insects are really refined predators, which usually, on growing to their full stature, kill the hosts they have been carefully sparing and, one might say, using as food-getting instruments. Since this is not exactly the form of parasitism exhibited by other organisms, such as the tape-worms, certain barnacles and bacteria, I prefer to call it "parasitoidism."

Yet even among the insects there are so many kinds and degrees of dependence on other insects that a concise classification is impossible. The phenomena are extremely diverse and protean, merging and melting into one another in the most bewildering manner. My limited time and the exigencies of exposition therefore compel me to condense and schematize. I am, moreover, dealing with a small fragment of a vast subject. The whole organic world is burdened with parasitism, so shackled and impeded by it that progressive evolution becomes inhibited in every group in which it appears, and the classes that have escaped its paralyzing touch are very few. Professor J. M. Clarke [1] has shown that parasitism made its appearance in marine animals as early as the Cambrian and that it has kept recurring ever since, specializing, and leading to "degeneration" and thus robbing group after group of species of all hope of further progress. Although they may persist for ages they are doomed to extinction, and only the independent forms, those that neither lapse into parasitic habits nor waste their vitality in nourishing parasites, stand any chance of becoming the ancestors of future types. We therefore belong to a lineage which, by some rare good fortune, escaped all the *culs-de-sac* of parasitism — till we became social.

The very conditions of social life tend to facilitate the

development of the host and parasitic relations. Not only
do the members of a society become more tolerant of alien
organisms in their midst and even domesticate and breed
them, but the nests and domiciles because of the protection
they afford, their higher temperature, the stores of food,
the refuse even, the helpless young and infirm old they contain — all representing so much nourishment — attract
hordes of predators, scavengers, inquilines, guests and parasites in the strict sense of the word. And the crowding
together of the social organisms greatly facilitates the interchange of all kinds of small parasites, such as mites,
molds and bacteria from host to host. On the other hand,
the members of a society are themselves normally temporary parasites of one another, the young of the adults, the
old of the young, and even the whole colony, as a unit, may
become a temporary or permanent parasite on the colony
of some other species. We noticed cases of this kind among
the social wasps and bees, namely *Vespa arctica* and *austriaca* and the various species of Psithyrus, and we shall
find more numerous examples among the ants. I did not
have time even to enumerate the alien beetles, flies, etc.,
that live in the nests of the social wasps and bees, but they
are numerous, and we shall find that the termites are sur-
passed only by the ants in the number of their parasites.

Although man furnishes the most striking illustrations of
the ease with which both the parasitic and host rôles may
be assumed by a social animal, his capacities in these directions have been little appreciated by the sociologists. Massart and Vandervelde [2] seem to be the only authors who
have attempted to do justice to the matter. Our bodies,
our domestic animals and food plants, dwellings, stored
foods, clothing and refuse support such numbers of greedy
organisims, and we parasitize on one another to such an

extent that the biologist marvels how the race can survive. We not only tolerate but even foster in our midst whole parasitic trades, institutions, castes and nations, hordes of bureaucrats, grafting politicians, middlemen, profiteers and usurers, a vast and varied assortment of criminals, hoboes, defectives, prostitutes, white-slavers and other purveyors to antisocial proclivities, in a word so many non-productive, food-consuming and space-occupying parasites that their support absorbs nearly all the energy of the independent members of society. This condition is, of course, responsible for the small amount of free creative activity in many nations. Biology has only one great categorical imperative to offer us and that is: Be neither a parasite nor a host, and try to dissuade others from being parasites or hosts. Of course, this injunction is no more easily obeyed than Kant's famous imperative, of which it embodies the biological meaning, for a parasite always treats its host as a means and not as an end, and the thoroughly parasitized host must abandon all hope of being an end in itself.

I have expressed myself somewhat drastically on human parasitism. If I attempted to utter all my opinions on the subject I should probably not be permitted to survive till the next lecture, even in so tolerant a community as Boston. But so vividly are the development and consequences of biological dependence illustrated by the ants that by confining myself to them, and possibly allowing a hint to escape here and there, you will be able to construct your own analogies. The more striking relations of ants to other organisms are enumerated in the accompanying list. I considered the relations to the Phytophthora (Fig. 78) in the preceding lecture, and our knowledge of the relations to the higher plants is in a state too controversial to admit of satisfactory exposition within the limits of this lecture.[3]

We may therefore confine our attention to social parasitism, or the behavior of ants as parasites and hosts of one another and to the myrmecophiles, or animals that use the ants as hosts. Social parasitism is exhibited by two series of species, one in which the parasitic and host colonies occupy separate though contiguous nests and therefore rear

Fig. 78. — Ants attending aphids on the roots of grasses and other herbs. (From a drawing by T. Carreras, after E. Step.

their broods in separate chambers, or nurseries, the other in which the two colonies have become so intimately united that they occupy a single nest and bring up their young in common.[4] It will be seen that not only each of these series, but also that of the myrmecophiles begins in predatory (indicated by asterisks) and terminates in definitively parasitic relations.

RELATIONS OF ANTS TO OTHER ORGANISMS

I. SOCIAL PARASITISM (Ants as Parasites)

    A. *Compound Nests* (Broods reared separately)
        *1. Brigandage (Cleptobiosis)
        *2. Thievery (Lestobiosis)
        3. Neighborliness (Plesiobiosis)
        4. Tutelage (Parabiosis, Phylacobiosis)
        5. Hospitality (Xenobiosis)

    B. *Mixed Colonies* (Broods reared together)
        *1. " Slavery " (Dulosis)
        2. Temporary Social Parasitism
        3. Permanent Social Parasitism

II. MYRMECOPHILY (Ants as Hosts)

        *1. Persecuted Intruders (Synechthrans)
        2. Indifferently Tolerated Guests (Synœketes)
        3. Mess-mates (Commensals)
        4. True Guests (Symphiles)
        5. External Parasites (Ectoparasites)
        6. Internal Parasites (Entoparasites)

III. TROPHOBIOSIS (Relations of Ants to Phytophthora, etc.)

IV. PHYTOPHILY (Relations of Ants to Plants)

The great armies of the nomadic legionary ants to which I alluded in my previous lecture often attack the nests of other ants and carry away and devour all their larvæ and pupæ. This is, of course, pure predatism and is not included in the list because it is hardly a true interspecific association. This is obviously prevented by the itinerant and highly carnivorous behavior of the plunderers. In the compound nests, however, the colonies of the two species occupy stationary nests which are so close together that their galleries may interdigitate or intercommunicate and permit one of the species to enter the nest of the other. Different ant colonies even of the same species are so hostile that their mere existence in such contiguity implies that one of the species is to some extent exploiting the other. That the manner of exploitation differs in different ants will be seen from the following brief account of the various known types of compound nests:

1. Certain small but aggressive ants, which secure at least a portion of their sustenance by waylaying the foraging workers of another species and snatching away their food, deserve the name of brigands. Such ants naturally

Fig. 79. — Mound of agricultural ant (*Pogonomyrmex occidentalis*) bearing a crater (at *a*) of a small brigand ant (*Dorymyrmex pyramicus*).

make their nests near those of the species they plunder. Thus *Dorymyrmex pyramicus* in our southwestern states often constructs its nests in the clearing surrounding or even on the large mounds, of harvesting ants of the genus Pogonomyrmex (Fig. 79).

2. In cases of what I call " thievery " the exploitation is more subtle and efficient. The thief-ants, all of which are subterranean and have very small workers, nest in the

earthen walls of populous ant or termite nests, much as the little red house ant (*Monomorium pharaonis*) nests in the walls of our dwellings. The chambers of the two nests are connected by extremely tenuous galleries, excavated, of course, by the thief-ants and permitting them to invade the nests and feed on the brood of their large neighbors, but preventing the latter from entering the nests of the robbers who are either ignored or overlooked on account of their diminutive size, and therefore carry on their depredations unhindered. The abundance of food which they thus secure enables them to rear very large queens and males, but the workers themselves are condemned to perpetual dwarfhood by their criminal mode of life. The most remarkable thief-ants are found in the large termite nests of the tropics, and the conditions described attain their most extreme expression in the genus Carebara. The workers are minute, pale yellow and blind, the queens and males deeply colored and several thousand times as large as the workers. Arnold has recently suggested that these extraordinary differences in size must make it impossible for the young queen to feed her first brood of workers and hence to establish her formicary in the typical independent manner of other ants. For this reason, when she leaves the parental nest to take her nuptial flight, she carries, attached by their mandibles to the tufted hairs on her feet, several workers, which thus accompany her till she has made her cell in some termite mound, and then take charge of rearing her first brood. On reading Arnold's account I examined a number of females and males of the Ethiopian *Carebara vidua* in my collection and at once found the minute workers attached as he describes. The accompanying sketch (Fig. 80) shows one of the queens carrying two workers. These, of course, also attach themselves to the males that leave the nest at the same time, but as they do not accompany the nest-founding queens and die just after mating, the workers that happen to choose air-planes of the wrong sex also perish.

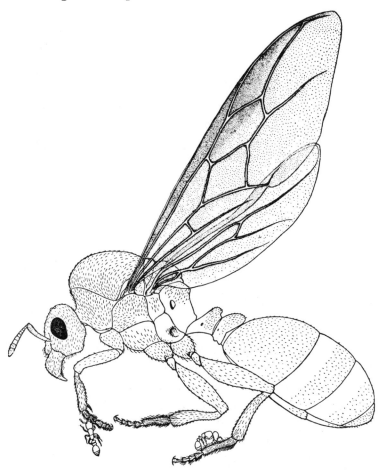

Fig. 80. — Winged queen of *Carebara vidua* carrying the minute, blind workers of her own species attached to her tarsal hairs.

3. What, for lack of a better term, I have called " neighborliness," is a very common relationship between two, or

more rarely three or even four species of ants living in nests, often with interdigitating but never with intercommunicating galleries, under the same stone or in the same log. Usually the ants of the different colonies, when they happen to meet, are more or less hostile. If one of the species is small and weak it undoubtedly derives some protection from merely living near a large and powerful neighbor, or the feebler may feed to some extent on the refuse of the larger form. When both species are large and aggressive they may perhaps find it advantageous to present a combined hostile front to the attacks of other ants.

4. What Forel calls " parabiosis," a word I have translated as " tutelage," seems to be a more definite relation of mutual or unilateral protection. In a typical case which I recently observed in British Guiana, we have two species, a small black Crematogaster (*C. parabiotica*) and a large brown Camponotus (*C. femoratus*) together inhabiting a large ball of earth which they build up around the branch of a tree. In this ball, which Ule calls an "ant-garden," because it supports numerous epiphytes, the Crematogaster inhabits the superficial, the Camponotus the central portions. When it is slightly disturbed the little black ants rush out to attack the intruder, but a more serious disturbance of the nest at once brings out the battalions of the much larger and extremely vicious brown species. The Crematogaster seem therefore to act as a skirmishing line for the Camponotus. Though the galleries of both species open freely into one another, and though the workers of both forage in long common files on the surrounding vegetation, they nevertheless keep their broods rigidly separated. The tutelary, or parabiotic relation, is evidently more mutualistic, or coöperative than any of the foregoing cases of the compound nests.

5. An interesting series of small species includes the
" guest ants " which live in still more intimate relations
with other species. One of the best examples is *Leptothorax emersoni* which I first found many years ago associated with the considerably larger *Myrmica canadensis* in
bog-like situations on our higher New England hills. The
Leptothorax inhabit small chambers at the surface of the
Myrmica nest and connect them by means of tenuous galleries with the chambers of their neighbors. The Leptothorax workers spend much of their time in the Myrmica
nest where they mount the backs of the workers and assiduously lick their bodies and especially their heads and
mouthparts. The Myrmicas seem greatly to enjoy this
performance and from time to time reward their little
guests with a droplet of regurgitated food. But while the
Leptothorax arrogate to themselves the right to mingle
freely with the Myrmicas and to flatter them into regurgitation, they resent the intrusion of the Myrmicas into their
own habitations and insist on bringing up their own brood
in perfect seclusion. Under natural conditions the Leptothorax are never seen to take any food, except from the
surfaces and crops of their hosts, but if kept for
some time by themselves in an artificial nest, they learn to
eat honey and insects like other ants. And if both species
are kept together in a glass nest without earth and therefore without materials for making separate chambers, the
Leptothorax eventually though very reluctantly permit the
Myrmicas to mingle the broods of both species and a true
mixed colony is formed.

The ants that live in the various compound nests are not
closely allied but belong to different genera or even subfamilies, a fact which may help to explain why they occupy separate nests and do not bring up their broods in

common, for the rearing of the brood is a very delicate operation and would be apt to differ considerably in unrelated species. We may therefore be prepared to find that mixed colonies are formed only by closely allied species, *i.e.*, either by those belonging to the same genus or to closely allied genera, and this proves to be the case. But before considering the various types of mixed colonies, two facts must be emphasized: First, many ants are fond of kidnapping the larvæ and pupæ belonging to other colonies, of their own or allied species. Frequently these kidnapped young are devoured, but in well-nourished colonies they may be permitted to complete their development and the emerging workers may be adopted as *bona fide* members of the colony, even if they belong to a different species. It is therefore possible to produce a mixed colony artificially by giving a colony the mature brood of some other species. In this manner Miss Fielde succeeded in inducing species belonging even to very different subfamilies to live together in perfect amity. It is also interesting to observe that ants thus reared in the colony of an alien species may be very hostile to their own sisters that have been left to grow up in the parental nest. Second, the mixed colonies found in nature are not in the first instance produced by a mere kidnapping of the brood of an alien species, but by the young queen of a parasitic species that is unable to start a colony independently, invading the nest of another species, which then becomes the host. The behavior of the invading parasite and the host colony differ in different species, but in nearly all the observed cases the host queen, if present, is eventually killed and her place is taken by the alien intruder. Since the queen ant is really the reproductive organ of the colony considered as a superorganism, the host colony may be said to be castrated and its sterile worker personnel is constrained to devote all its energies to rearing the brood which is forthwith produced by the fecund parasite. With these general statements in mind we may turn to the three types of mixed colonies, those of the slave-makers, the temporary and permanent social parasites:

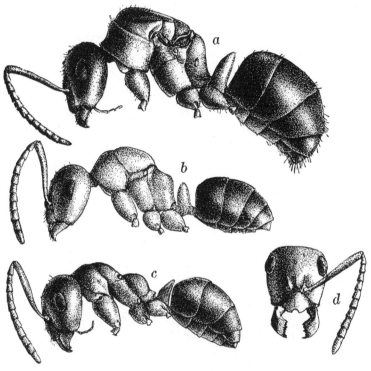

Fig. 81. — Blood-red slave-maker (*Formica sanguinea*) in profile; *a,* queen, with wings and legs removed; *b,* pseudogyne; *c,* worker; *d,* head of same from above, showing the characteristic notch in the clypeus.

1. The peculiar phenomena known as slavery, or dulosis, which occur in two genera of Formicinæ, Formica and Polyergus, and two genera of Myrmicinæ, Strongylognathus and Harpagoxenus, represent three phylogenetic stages, a primitive stage in *Formica sanguinea,* a culminating stage

in Polygergus and a degenerate or evanescent stage in Strongylognathus and Harpagoxenus. The "blood-red slave-maker" *F. sanguinea* (Fig. 81), is a common but rather local red ant, with black or brown gaster and is represented by numerous subspecies and varieties ranging over northern Europe, Asia and North America. It can be readily distinguished from the other species of the genus, at least in the Old World and the eastern United States, by the pronounced notch in the clypeus, or small shield at the anterior end of the head. The worker and queen look as if they were hare-lipped. *Sanguinea* is one of the most intelligent of ants and therefore one of the most interesting to keep in artificial nests. Its habits were first studied more than a century ago by Pierre Huber, the son of the blind François Huber, and have ever since commanded the attention of myrmecologists, because its armies of workers make periodical forays on the colonies of the common black *Formica fusca*, carry the worker larvæ and pupæ into their nest and permit many of them to emerge and become members of their colony. Thus the colony is mixed, and the black individuals, on account of their color and provenience, have been called "slaves." It is evident, however, that this term is inappropriate, for a slave is "a man who is the property of another, politically and socially at a lower level than the mass of the people, and performing compulsory labor" (Nieboer), and none of these distinctions applies to the *fusca* workers in the *sanguinea* nest. They are more properly called "auxiliaries" (Hilfsameisen), but I shall use the old term on account of its brevity. At least one of the subspecies of *sanguinea* (*aserva*) does not make slaves, and the colonies of some of the other forms give up the habit after a time, for the *sanguinea* colony, when once established, is quite able to lead an independent life. Darwin and others offered various explanations of the peculiar slave-making habit of *sanguinea,* but its meaning remained obscure till 1904 when I found that it had its origin in the behavior of the young queen. She is quite unable to found a colony independently and therefore, after her marriage flight, may adopt one of three courses: she may return to the nest in which she was reared or enter some other *sanguinea* nest, or she may invade a nest of *F. fusca.* As the first and second courses are sometimes adopted by other ants and do not lead to the formation of mixed colonies, they need no further consideration in this place and we may confine our attention to the last. As soon as the *sanguinea* queen invades a *fusca* colony, she becomes greatly excited and interested in the brood, seizes and collects in a small pile as many pupæ as she can snatch up and mounts guard over them. She slays any *fusca* workers that are bold enough to attempt to regain their property and is therefore soon left in undisputed possession of her plunder. Eventually *fusca* workers emerge from the cocoons and at once assume a friendly attitude towards the queen, feed her by regurgitation and behave towards her as if she were their own mother. She begins to lay eggs and the resulting larvæ are fed and reared by the black workers, so that when the *sanguinea* emerge a mixed colony is established. These workers show that they have inherited their mother's proclivities by kidnapping the brood of neighboring *fusca* colonies, but they do this as an army and carry the *fusca* brood to their nest. In some colonies, as I have stated, this kidnapping, or slave-making proclivity may disappear after a time, and in *aserva* it seems to disappear very early or perhaps is not even inherited by the workers. In such cases, therefore, the personnel of old colonies may be made up entirely of *sanguinea* after the batch of *fusca* workers kidnapped and reared by the queen has died of old age. It is evident that slavery is at bottom a form of predatism and has its origin in the inability of the young queen to establish a colony without the aid of workers. Unlike the great majority of ant-queens, she has been unable to store enough food in her body to stand the strain of long fasting and nourishing her first brood. In another sense she is, of course, a parasite and the *fusca* workers represent the host. Owing to the fact that the colony may eventually cease to increase its worker personnel by the kidnapping of *fusca* brood, we may call this type of slavery temporary, acute, or faculative.

The species of Polyergus, or " amazons," as they were called by Pierre Huber, have much the same distribution as *sanguinea* and have the same species of Formica as slaves, but their method of securing the latter is more highly perfected. The amazons are very beautiful red ants (except the Japanese *P. samurai*, which is black), and their mandibles are slender and sickle-shaped and perfectly adapted to fighting but of no use for digging in the earth or capturing food (Fig. 82). Hence these insects are unable to make nests or even to feed themselves or care for their own young, but are absolutely dependent on their slaves. Like *sanguinea* the amazons make periodical forays, which for some unknown reason are always carried out in the afternoon, but their armies show a more perfected tactical organization and the subjugation and plundering of the *fusca* colonies are effected with much greater dispatch and precision — one might say with the most consummate *éclat*. At the approach of the amazons the *fusca* workers usually flee in dismay, but if they offer any resistance the amazons pierce their heads with the sickle-shaped man-

dibles. The young on emerging from the kidnapped pupæ
excavate the nest, feed the Polyergus and bring up their
brood but do not accompany the armies on their raids.
The initial stages in founding the colony have been studied
by Emery, who found that the young Polyergus queen
secures adoption in some small, weak *fusca* colony after
killing its queen by piercing her head. She then produces

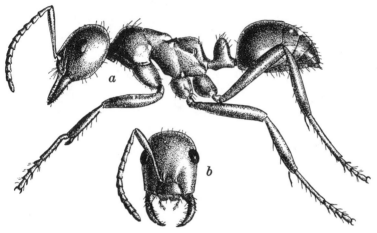

Fig. 82. — *a,* Worker of *Polyergus lucidus,* the " shining amazon," a
permanent slave-maker of the Eastern United States; *b,* head of same,
showing the sickle-shaped mandibles.

her brood which will later make the slave-raids on the
fusca colonies. Since this raiding proclivity never lapses
even in old colonies, Polyergus is to be regarded as a
chronic, or obligatory slave-maker. An amazon crimson on
a field sable with the device " *stultus sed pugnax* " might
be an appropriate coat-of-arms for some of the military
castes that have flourished during the course of human
history.

In Europe there are several species of the interesting
genus Strongylognathus (Fig. 83*a-c*), which have sickle-

shaped mandibles like the Polyergus and always live in the colonies of the common pavement ant, *Tetramorium cæspitum* (Fig. 83d).   Our fragmentary knowledge indicates that we have here some of the degenerate or evanescent stages of slavery.   The workers of *S. rehbinderi* and *huberi* seem still to make forays on Tetramorium colonies and to carry home their brood, and Kutter has recently shown that *S. alpinus,* a form I discovered some years ago near the head-waters of the Visp, within sight of the Mat-

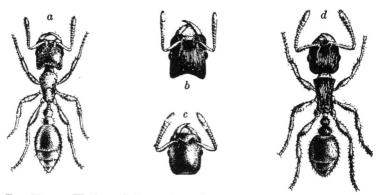

Fig. 83. — *a,* Worker of *Strongylognathus testaceus,* a degenerate slave-maker of Europe; *b,* head of female of same; *c,* head of worker of *S. huberi* an allied species; *d,* worker of the pavement ant (*Tetramorium cæspitum*), the host of the species of Strongylognathus.

ternorn, makes nocturnal slave-raids and is accompanied by its slaves, which do most of the fighting and carry home the brood of their own species.   In this case the slaves are really the masters and seem to use the Strongylognathus merely as a means of disconcerting or terrifying the colonies of *cæspitum* whose brood they are bent on kidnapping.   Finally, *S. testaceus,* the best-known species of the genus, no longer makes forays and is tending to lose its worker caste.   Wasmann, Mrázek, Forel and I have found that colonies of *cæspitum* infested by this species may retain the host queen. In order to establish her colony, therefore, the young *testaceus* queen probably associates herself with a young, nest-founding *cæspitum* queen. In the mixed colonies of other species of Strongylognathus the host queen appears to be eliminated as in the colonies of Polyergus.

2. In 1904 I detected another method of forming mixed colonies, which I called temporary, although I might have called it acute, social parasitism. It is practiced by a number of ants, especially by several North American species of Formica that have unusual queens. In some species they are peculiarly colored or furnished with long yellow hairs, in others they are extremely small, smaller even than the largest workers (Fig. 84). The young queen of these ants enters the nest of another Formica belonging to the *fusca* or *pallide-fulva* group and is very apt to be adopted, probably on account of her smaller size or other physical attractions. The fate of the host queen in such invaded nests has not been ascertained but she is probably killed by her own workers. The parasite then proceeds to produce her brood, which is reared by the host workers, and a mixed colony results. As there is no inclination on the part of the queen's offspring to plunder other nests of the host species, and as all the host workers die off in the course of a few years, a pure colony of the parasitic species is left behind and may grow to be very populous and aggressive, without showing any signs of its parasitic origin — a beautiful analogue of some human institutions, which after starting in humble and cringing parasitism have come to acquire during the centuries a most exuberant and insolent domination. Our common mound-building ant (*Formica exsectoides*) is one of these successful temporary parasites which starts its opulent colonies with the aid of the ubiquitous *F. fusca*

var. *subsericea.* Since my observations were published several European Formicas, including the well-known mound-building *rufa,* and ants of certain other genera (Lasius, Bothriomyrmex, Aphænogaster, etc.) in various parts

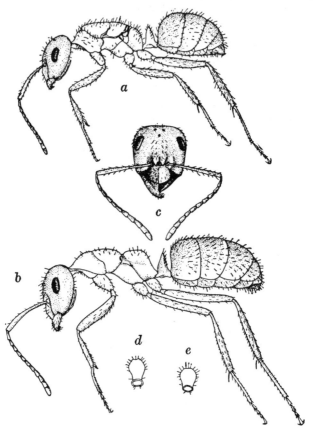

Fig. 84.— A temporary parasite (*Formica microgyna*). *a,* queen, with wings removed; *b,* large worker drawn to same scale; *c,* head of same; *d* and *e,* petiole of worker and queen seen from behind.

of the world have been found to be temporary social parasites. One of the most interesting of these is the Dolichoderine *Bothriomyrmex decapitans* which Santschi observed in Tunis. The young queen, on descending from her marriage flight, wanders about on the ground till she finds the nest of a *Tapinoma nigerrimum* colony, when she permits herself to be seized and " arrested " by its workers. These then proceed to drag her into their burrow by her legs and antennæ. After entering the nest the parasite may be attacked from time to time by the workers, but she takes refuge on the brood or on the back of the larger Tapinoma queen. In either of these positions she seems to be quite immune from attacks, probably because her own odor is overlaid by that of the brood or the host queen. Santschi observed that the parasite often spends long hours on the back of the Tapinoma queen and that while in this position she busies herself with sawing off the head of her host! By the time she has accomplished this cruel feat, she has acquired the nest-odor and is adopted by the Tapinoma workers in the place of their unfortunate mother. The parasite thereupon proceeds to keep them busy bringing up her brood. They eventually die of old age and the nest then becomes the property of a thriving, pure colony of *Bothriomyrmex decapitans*.

3. There are more than a dozen genera of ants from various parts of the world, which may be classed as permanent, or chronic social parasites. They have all completely lost the worker caste so that in this respect they closely resemble the parasites among the social wasps and bees. The young queens enter the nests of other ants and secure adoption, like the queens of the temporary social parasites. The host queen seems to be regularly assassinated by her own workers. At least this has been observed by Santschi in the case of *Wheeleriella santschii,* which lives in the nests of the common North African *Monomorium salomonis.* After fecundation the *Wheeleriella* queen roams

about over the surface of the soil in search of a Monomorium nest. When near the entrance of one of them she is
" arrested," to use Santschi's expression, by a band of Monomorium workers, which tug at her legs and antennæ and
draw her into the galleries. Sometimes she may be seen to
dart suddenly into the entrance of her own accord and is
arrested within the nest. There are no signs of anger on
the part of the Monomorium, and she is soon permitted to
move about the galleries unmolested. The workers then
begin to feed and adopt her and in the course of a few days
she lays her first eggs, which are accepted and cared for by
the host. The parasite pays no attention to the much
larger Monomorium queen, but the latter is eventually assassinated by her own workers. Other species, like the
famous *Anergates atratulus* (Fig. 85) of Europe and the recently discovered *Anergatides kohli* of the Congo, are much
more highly modified and represent the last stages of parasitic degeneration. In the former, which lives with *Tetramorium cæspitum,* the queen is small and winged (Fig.
85a), but after deälation and adoption her gaster swells
enormously with eggs till she resembles an old termite
queen (Fig. 85b). The male (Fig. 85d) is wingless and
pupa-like and unable to leave the nest so that mating takes
place between brothers and sisters ("adelphogamy" of
Forel). The conditions in Anergatides, which is a parasite
of *Pheidole melancholica,* are somewhat similar. In the
workerless parasites the offspring of the intrusive queen
are, of course, all males and females and are produced
during the life-time of the host workers. The colonies are
therefore mixed throughout their existence which is necessarily terminated by the death of the host.

While all myrmecologists now agree in recognizing the
three types of social parasitism and their origin in the be-

havior of the young queens, there is still disagreement in regard to their phylogenetic derivation. I at first believed that they had all had their inception in the passive adoption of insufficiently endowed, young queens by colonies of their own species, and Wasmann has consistently adhered to this view. Emery and Viehmeyer, however, see in the aggressive, predatory behavior of the *sanguinea*

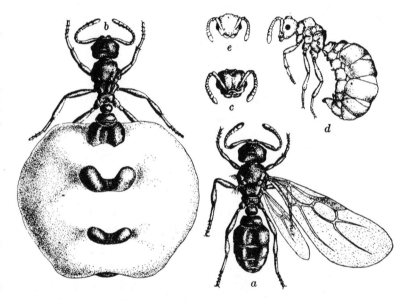

FIG. 85. — A workerless, degenerate, permanent social parasite (*Anergates atratulus*) of Europe. *a*, virgin queen; *b*, old, egg-laying queen with enlarged gaster; *c*, head of same from front; *d*, male, which is wingless and pupoidal; *e*, head of same.

queen a stage from which temporary and permanent social parasitism may be more naturally derived. I am now inclined to believe that these investigators are nearer the truth and that the adoption of queens by colonies of their own species is a distinct phenomenon, which may readily lead to the formation of new colonies by a kind of swarming, analogous to that of many social bees and wasps, but not to the series of parasitic developments which we have been considering.

All the parasitic ants are rare or local. The permanent, or chronic social parasites, especially, are so very scarce that they must be on the very verge of extinction — a fact which shows that parasitism, so far as the race is concerned, is anything but a promising or profitable business. But even the individual parasite buys its rare successes very dearly, for it must often run the gauntlet of great resistance and animosity on the part of a too healthy host and must at the same time carefully avoid seriously injuring that host and thus bringing about its own destruction. Parasitism in the queen ant may, indeed, be regarded as a kind of compensation or overcompensation for her inability to rear a brood of workers. One is reminded of the overcompensations (megalomania) resorted to by some human beings with pronounced inferiority complexes. Was it an inkling of this that led the ancients to make Hercules the tutelary deity of parasites? That the parasitic queen's inability has been acquired and fixed during the past history of the species is suggested by the singularly close genetic relation of the parasites to their hosts. In the great majority of cases, as indicated in the accompanying list, the parasite belongs either to the same genus as its host or to a genus descended from that of its host. This is equally clear from detailed lists of the other parasitic ants, wasps and bees and shows that the parasite originally led an independent life but took to exploiting some common, allied species of the same genus.[5] Probably the exploitation was at first predatory as it still is in certain Psammocharid wasps and *Formica sanguinea,* because food was more expeditiously secured by such tactics. In the

course of time the parasite's adaptations to its host became
increasingly refined and were reflected in its structure as

PERMANENT SOCIAL PARASITES (ANTS WITHOUT WORKERS)

PARASITES	HOSTS	HABITAT	ANCESTRAL GENUS
Sympheidole elecebra	Pheidole ceres	Nearctic	Pheidole
Epipheidole inquilina	Pheidole ceres	Nearctic	Pheidole
Parapheidole belti	(?) Pheidole sp	Malagasy	Pheidole
Sifolinia laurœ	(?) Pheidole sp	Palearctic	Pheidole
Anergatides kohli	Pheidole melancholica	Ethiopian	Pheidole
Wheeleriella santschii	Monomorium salomonis	Palearctic	Monomorium
Wheeleriella adulatrix	Monomorium subnitidum	Palearctic	Monomorium
Wheeleriella wroughtoni	Monomorium indicum	Palearctic	Monomorium
Epœcus pergandei	Monomorium minimum	Nearctic	Monomorium
Epixenus andrei	Monomorium venustum	Palearctic	Monomorium
Epixenus biroi	Monomorium creticum	Palearctic	Monomorium
Myrmica myrmoxena	Myrmica lobicornis	Palearctic	Myrmica
Hagioxenus schmitzi	Tapinoma erraticum	Palearctic	Monomorium
Anergates atratulus	Tetramorium cœspitum	Palearctic	(?) Tetramorium
Pseudoatta argentina	(?) Mœllerius balzani	Neotropical	Mœllerius
Plagiolepis nuptialis	Plagiolepis custodiens	Ethiopian	Plagiolepis

generic distinctions, as we see, e.g., in the species of Polyergus, which are obviously modified Formicas. The descent to Avernus became steeper and more slippery, as the
parasite, yielding to inertia, became chronically and ab-
jectly dependent on its host and condemned itself to physiological and numerical inferiority ("misère physiolog-
ique"). The next stage is extinction, after a longer or
shorter period of hopeless specialization ("degeneration").
This or a very similar story has been so often repeated in
all the classes of the animal and plant kingdoms that the
number of forms which during geological time have descended to the limbus parasitorum must be considerable.

Having considered the ants as parasites and hosts of
one another, we may now turn to the cases in which they
act as the hosts of insects belonging to very different orders,
the myrmecophiles, or ant-guests. Here we enter on a

vast and very intricate subject to which I shall be unable to do justice in the short time at my disposal.[6] Fully 2,000 species of myrmecophiles have been described, and no doubt the number will be more than doubled when the nests of the many species of tropical ants have been carefully explored. The myrmecophiles include not only members of nearly all the different orders of insects but also many spiders, mites, millipeds and land-crustaceans — a weird, one might almost say demoniacal horde of creatures, which have been induced to live in more or less intimate and maleficent relations with the ants by the obvious advantages of the association. Ants' nests furnish admirable hiding or lurking places and are at night and during the winter months somewhat warmer than the surrounding soil. They often contain quantities of food or refuse, and the helpless brood, callow and injured ants may be stealthily devoured. Furthermore, the ants may be wheedled into adopting and feeding alien insects, as if they were their own young or ants of the same species. All the forms of exploitation, therefore, from predatism, adoption and domestication to external and internal parasitism have been developed by the myrmecophiles. It may be said that these creatures have searched out and taken advantage of every vulnerable point in the ant's structure and behavior, just as every human idiosyncracy, frailty and virtue has been exploited by some cunning human parasite.

There are two reasons why we must consider the myrmecophiles in these lectures. First, many of them live only with particular ants and really form a constituent though not an essential part of their colonies, for although they are not present in all colonies, they can not exist apart from the ants except while migrating from one nest to another. They are, in fact, a more integral component of the colony in which they occur than are the domestic animals in the human community. Many of our domestic animals are still able to return to an independent, feral life, but this is impossible for the more typical and highly specialized ant-guests. Second, the ant-guests afford a very striking, indirect or pathological demonstration of the extraordinary intensity of the brood-nursing propensities of ants. Any insect possessed of the glandular attractions, which I shall presently describe, can induce the ants to adopt, feed and care for it and thus become a member of the colony, just as an attractive and apparently well-behaved foreigner can secure naturalization and nourishment in any human community. But the procedure among the ants is more striking, because the foreigners are so very foreign, that is, belong to such alien and heterogeneous groups. Were we to behave in an analogous manner we should live in a truly Alice-in-Wonderland society. We should delight in keeping porcupines, alligators, lobsters, etc., in our homes, insist on their sitting down to table with us and feed them so solicitously with spoon-victuals that our children would either perish of neglect or grow up as hopeless rhachitics.

Although every species of myrmecophile has its own methods of securing food and lodgings in the ant nest I shall describe only a few examples to illustrate the exploitation of the trophallactic habit.

1. There are probably several myrmecophiles that steal the food given to the larva, but the only case that has been adequately described is the larva of a small fly, *Metopina pachycondylæ,* which I found many years ago in Texas infesting the nests of a Ponerine ant, *Pachcondyla montezumia* (Fig. 86).[7] This ant feeds its larvæ in a very primitive manner with pieces of insects and thus exposes itself

to the inroads of the Metopina. Its small larva clings to the neck of the ant larva by means of a sucker-like posterior end and encircles its host like a collar. Whenever the ant larva is fed by the workers with pieces of insect placed on its trough-like ventral surface, within reach of its

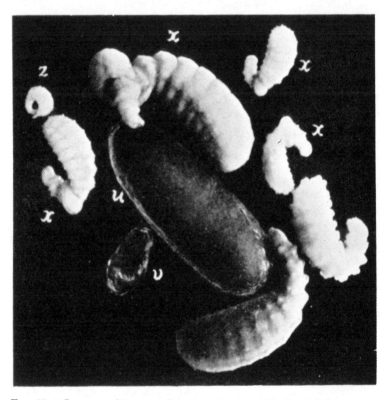

Fig. 86. — Larvæ and pupæ of a ponerine ant (*Pachycondyla montezumia*) and its commensal (*Metopina pachycondylæ*). The Pachycondyla larvæ marked *x* have each a Metopina larva around the neck; *z*, isolated Metopina larva; *v*, Metopina puparium; *u*, cocoon of Pachycondyla.

mouthparts, the larval Metopina uncoils its body and partakes of the feast; and when the ant larva finally spins its cocoon it also encloses the Metopina larva within the silken web. The commensal, however, moves to the caudal end of its host and forms a small flattened puparium which is applied to the wall of the cocoon. This is obviously an adaptation for preventing injury from the jaws of the worker ants when the cocoon is being opened and the callow extracted from its anterior end. The ant hatches before the Metopina and the empty cocoon, with the puparium concealed in its posterior pole, is carried to the refuse heap. Here the fly emerges and escapes from the cocoon by the opening through which its host emerged. The Metopina consumes so little food and is so considerate of its host that it can hardly be said to produce any injurious effect on the colony; at any rate the larvæ which have borne commensals develop into perfectly normal workers. The ants clean the commensals when they are cleaning their own progeny and show no signs of being aware of their presence in the nest.

FIG. 87. — *Atelura formicaria* about to snatch the droplet of food that is being regurgitated by one *Lasius mixtus* worker to another. (After C. Janet.)

2. *Lepismina* (*Atelura*) *formicaria* (Fig. 87) is a small, primitive insect which lives in the nests of *Lasius mixtus*. Its body tapers rapidly behind and is covered with slippery scales so that it is not easily caught by the ants. It

is, moreover, extremely agile and circumspect, because it has not succeeded in ingratiating itself with its hosts. Janet,[8] after providing a colony of *Lasius mixtus* with honey was able to make the following observations on the behavior of the insect. " From the instant that the first foragers returned to the nest, the Lepismina showed by their excitement that they perceived the odor of honey. Soon a considerable number of ants were grouped in couples for the purpose of regurgitating. They elevated their bodies slightly and often raised their fore legs, thus leaving a vacant space under their heads. As soon as a Lepismina came near such a couple, it thrust itself into the space, raised its head, suddenly snapped up the droplet that was passing in front of it and made off at once as if to escape merited pursuit. But the ants standing face to face are not free enough in their movements even to threaten the audacious thief, who forthwith proceeds to take toll from another couple and continues these tactics till his appetite is appeased."

3. A more subtle method of obtaining regurgitated food is adopted by the large mites of the genus Antennophorus (Fig. 88), which have been studied by Janet, Wasmann, Karawaiew and myself.[9] These mites, which have conspicuously long fore legs and attach themselves to the bodies of the workers, whether present in odd or even numbers, always orient themselves in a symmetrical position with respect to their host. When only one Antennophorus is present it clings to the gula, or chin of the ant, with its fore legs directed toward the ant's mouthparts. When two are present, there is one on each side of the head or one on each side of the gaster; in the former case the antenniform appendages are directed towards the anterior, in the latter towards the posterior end of the ant's

body. When there are three mites, one attaches itself to the chin and the two others to the sides of the gaster. Four place themselves in pairs on the sides of the head and gaster. If six are present, which rarely happens, four are arranged in pairs on the sides of the head and gaster while

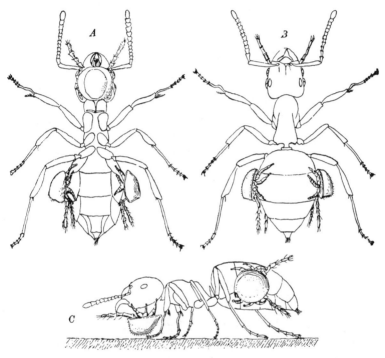

Fig. 88. — *Lasius mixtus* worker carrying three symmetrically oriented mites (*Antennophorus pubescens*). *A,* ventral; *B,* dorsal; *C,* lateral view. (After C. Janet.)

of the two remaining individuals, one attaches itself to the chin, the other to the mid-dorsal surface of the gaster. Janet believes that these symmetrical arrangements are for the purpose of balancing the burden and thus making it easier for the ant to carry. When attached to the head the mite obtains its food by drinking the regurgitated droplet as it is being passed to or from the mouthparts of the host, or it titillates the ant with its antenniform legs and induces her to regurgitate for its special benefit. The mites attached to the gaster obtain their food by stroking other ants in the vicinity or by reaching out and partaking of the droplets as they pass from one ant to another. The ants try to rid themselves of the parasites when they first attach themselves, but after they have taken up their definitive, symmetrical positions, they seem to be tolerated with indifference.

Most of the species of Antennophorus have been described from Europe, but I have found two species (*donisthorpei* and *wasmanni*) rather common near the Arnold Arboretum in Boston. They live in essentially the same manner as the European Antennophori with our small yellow ants of the genus Lasius, and its subgenus Acanthomyops. All these Lasii are hypogæic and devote themselves to attending snow-white plant-lice and mealy-bugs on the roots of our forest trees, and since the mites occur only with these ants it would seem that they, the plant-lice, the mealy-bugs, the mites and the forest trees are all so many members of a peculiar, subterranean association, or biocœnose. The plant-lice and mealy-bugs pump the juices out of the plants and pass on to the soliciting ants the unassimilated portions in the form of honeydew, some of which the ants regurgitate to the mites that ask them for it by aping with their long, hairy fore legs the antennary movements of the hungry ants. In other words, the ants serve as cup-bearers, distributing to one another and to the indolent, sedentary Antennophori the nectar which the tapster plant-lice and mealy-bugs keep drawing from their vegetable hosts. Owing to this interesting biocœnetic arrangement the worker Lasii do not have to come to the surface of the ground to seek their food. The eyes of the workers have therefore become so minute that their visual powers must have nearly or quite disappeared. Perhaps we can best appreciate the relations of the ants to the mites if we fancy ourselves blind, condemned to live in dark cellars and continually occupied with pasturing and milking fat, sluggish cows, yielding quantities of strained honey instead of milk. Then let us suppose that occasionally there alighted on our cheeks or backs small creatures which, by placing themselves in positions symmetrical to the median longitudinal axis of our bodies, took great care not to annoy us, and stretched forth to us from time to time small, soft hands, like those of our friends, begging for a little honey, should we not under the circumstances, treat these little Old Men of the Sea with much lenity and even with something akin to affection?

4. The behavior of the myrmecophiles I have been considering is simple and transparent compared with that of the true ant-guests or symphiles, which are really the *élite* of all the insects that live in ant colonies. They comprise several hundred species of beetles belonging to a number of natural families but showing a very singular, convergent agreement in certain characters, such as a deep, oily red color and peculiar tufts of golden yellow hairs, or trichomes on various parts of their bodies (Fig. 89). Their antennæ and mouth-parts, too, are in many cases peculiarly modified, the former for soliciting, the latter for receiving regurgitated food. The trichomes surround the openings of singular glands, the aromatic, volatile secretions of which flow along the hairs and are licked off by the ants. So inordinately fond are the ants of these secretions that they cherish the beetles, feed them and carry them to safety when the nest is disturbed or to new nests when the old ones

have to be abandoned. The beetles breed in the colonies and their larvæ are often treated with even greater solicitude than the ant larvæ.

Probably the most remarkable of these true guests are the Lomechusini, which have been studied by Wasmann for more than 30 years.[10] They belong to the rove-beetles, or Staphylinidæ and comprise only three genera: Lo-

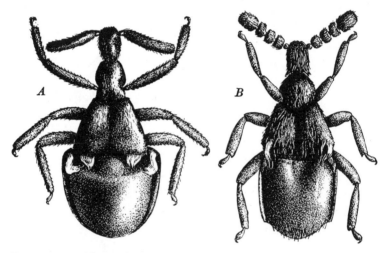

Fig. 89.—A, *Adranes lecontei* of North America and B, *Claviger testaceus* of Europe, two guest beetles, with golden yellow trichomes at the tips of their wing-cases and at the base of the abdomen.

mechusa (Fig. 90) and Atemeles, peculiar to Europe and Northern Asia, and Xenodusa (Fig. 91), known only from the United States and Mexico. The species of Atemeles and Xenodusa have two hosts, those of the former living during the summer and breeding in Formica colonies but hibernating in colonies of Myrmica, the latter also breeding with Formica but hibernating with our large carpenter ants of the genus Camponotus. Lomechusa, on the other hand, has only one host, *Formica sanguinea*, with which it

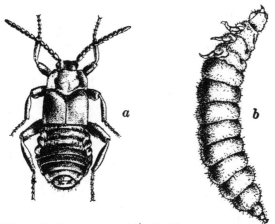

FIG. 90. — *a*, A European guest-beetle (*Lomechusa strumosa*) and *b*, its larva, which live with colonies of the blood-red slave-maker (*Formica sanguinea*).

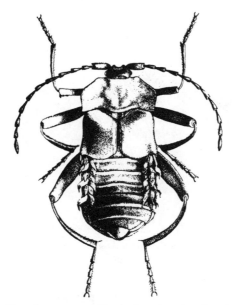

FIG. 91. — *Xenodusa cava*, a North American beetle which breeds in the nests of Formica during the summer and passes the winter in the nests of Camponotus. Note the tufts of trichomes along the sides of the abdomen.

lives throughout the year. The adult beetles of all three genera look much alike. They have long, mobile antennæ, short wing-cases and a voluminous abdomen, which can be curled up over the thorax and is provided on each side above with a segmental series of beautiful golden trichomes. Atemeles and Xenodusa beg their food from the ants by stroking their cheeks with the fore feet. Their larvæ are active, have long legs and employ the same method as the beetles in persuading the ants to regurgitate. They also devour the defenseless Formica larvæ. The adult Lomechusa is more passive in its behavior and uses its antennæ in soliciting food (Fig. 92). Its larvæ (Fig.

Fig. 92. — Worker of the blood-red slave-maker, *Formica sanguinea,* feeding its guest beetle, *Lomechusa strumosa.* (After H. K. Donisthorpe.)

90*b*) have very short legs and are unable to run about but lie among the ant brood. They eat the brood but are also fed by regurgitation. In all probability they secrete fatty exudates which are greatly appreciated by the ants.

At any rate, the ants seem to prefer the Lomechusa larvæ to their own, or perhaps regard them as unusually promising ant larvæ. In consequence of this infatuation the Lomechusa larvæ often destroy the greater part of the brood, so that in *sanguinea* colonies heavily infested with the parasites the queen larvæ develop abnormally. Either these larvæ are neglected, or the ants actually endeavor to convert them into workers, because they feel that this caste is inadequately represented in the colony. But whatever be the treatment of the queen larvæ, they develop into pathological adults, known as " pseudogynes " (Fig. 81c) — abortive creatures, resembling workers in size and in the shape of the head and gaster, but with a more voluminous and convex thorax, approaching that of the queen. They are paler than the normal workers and very lazy, cowardly and incompetent. Usually they constitute 5 to 7 per cent., less frequently 20 per cent. or more, of the personnel of an infested *sanguinea* colony. Their appearance in a nest indicates that the colony is in a diseased condition and on the road to extinction, as the result of Lomechusa infection. Similar pseudogynes are also produced in the Formica colonies infested with Atemeles and Xenodusa, but not in the Myrmica and Camponotus colonies in which the beetles hibernate, because they do not breed among their winter hosts and can not therefore interfere with the normal development of their brood.

Such being the effect on the colonies that harbor the Lomechusini, one naturally inquires, why the habit of rearing the parasites has not long since led to the extinction of the Formicas. This question has been partially answered by Wasmann. He finds that the ants treat the Lomechusa larvæ like their own, even when they are ready to pupate. *F. sanguinea,* like many other ants, buries its full-grown larvæ in the soil in order that they may spin their cocoons and pupate within them. After pupation the cocoons are unearthed, cleaned and stacked up in the chambers of the nest. Now the full-grown Lomechusa larvæ also need to be buried in order to pupate, though they do not spin cocoons, but they must not be unearthed after pupation, like the ant brood, or they perish. The ants, however, are utterly ignorant of these different developmental requirements and therefore unearth as many of the Lomechusa pupæ as they can find. Thus death in the guise of what might be called a regulatory Nemesis, overtakes all except the few pupæ that have been overlooked by the ants, but these few suffice to insure the survival of the species. More recently Wasmann has claimed that *sanguinea* takes a particular liking to certain pairs of beetles and eliminates the less attractive individuals from the colony. Of course, this would still further tend to reduce the incidence of parasitism and its baneful effects on the host.

The behavior of *sanguinea* and Lomechusa has seemed to Wasmann so unique and extraordinary that he has used it on every occasion as furnishing brilliant proofs of the absence of intelligence in ants, of the impotence of natural selection, of the possession by *sanguinea* of a singular, innate Lomechusa-fostering instinct, and therefore of a new type of selection, which he calls " amical selection," and, finally, even of the Divine Wisdom in maintaining the equilibrium in nature. Being a very accomplished Jesuit he has devoted many hundred pages and much scholastic casuistry to these "proofs." To the biologist who is under no compulsion to make his conclusions square with the philosophy of St. Thomas Aquinas, the behavior of *sanguinea* appears in a very different light, as a brilliant example of the perversion of appetites. Escherich has compared the infatuation of the Lomechusa-cherishing *sanguinea* with alcoholism in the human species. It might also be compared with a cat's infatuation with catnip, and the rearing of the Lomechusa larvæ by the ants would seem to be due to the same kind of instinct perversion that we observe in the birds that rear cuckoos, the occasional cat that rears a puppy or the hen that adopts kittens. It would seem to be no more necessary to postulate a special Lomeschusa-rearing instinct in *sanguinea* than a special ice-cream instinct in our children or a special Havana-cigar instinct in old bachelors.[11]

That the behavior of *sanguinea* towards Lomechusa proves nothing in regard to intelligence, even when the word is used, as it is always used by Wasmann, in the scholastic sense, that is, as equivalent to reason, or ratiocination, has been shown by Hobhouse.[12] After referring to Wasmann's conclusions, he says: " Difficult as it is to conceive the psychological conditions under which such contrasts are possible, we may still get some help from the analogy of human action. When comparative psychologists take occasional inconsistency as proving the utter absence of intelligence (in animals) they are using an argument which would equally disprove the existence of intelligence in man. After all, is an ant-nourishing parasite that destroys its young guilty of a greater absurdity than, say a mother promoting her daughter's happiness by selling her to a rich husband, or an inquisitor burning a heretic in the name of Christian charity, or an Emperor forbidding his troops to give quarter in the name of civilization? The mother really desires her daughter's happiness, but her conception of the means thereto is confused, and rendered self-contradictory by worldly ambitions. The inquisitor's conception of Christian charity is similarly corrupted by the subtle corporate egoism of a Church and the cruel pedantry of bad theology. Even the Emperor has some conception of civilization, but it is the civilization of militarism. In all cases there impinge on the avowed plan of action conflicting impulses of a kind not to stop the course of action, but to merge in it and distort it."

Hobhouse also comments trenchantly on Wasmann's conception of the manoeuvers of the Divine Wisdom in maintaining the equilibrium of parasite and host, but I will pass over this matter, because it involves a consideration of the problem of evil and belongs to the philosopher and the theologian. The cases of social parasitism considered in the first part of this lecture and the behavior of *sanguinea* and Lomechusa suggest a few concluding remarks on a matter that is often overlooked. It is evident that a parasitic species is more seriously affected and at a much greater disadvantage than a host species, notwithstanding the suffering or death that may be inflicted on individual host colonies. Without exception, the hosts of all the known social parasites and Lomechusini are very common, prolific, widely distributed and therefore very plastic, or adaptable ants and the instances of infection of these hosts are local or sporadic. This is even the case with *Formica fusca,* which occurs practically over the whole northern hemisphere and is infested by an extraordinary series of social parasites. It is enslaved by *F. sanguinea* and *Polyergus rufescens* and serves as the temporary host of many parasitic Formicas, including *rufa, truncicola, dakotensis, exsecta, exsectoides,* etc. We may conclude, therefore, that *fusca* and the other hosts of parasitic ants and Lomechusini have developed more or less resistance or even a certain local immunity to the inroads of the parasites, and that the parasites even when successful exploit merely that margin of super-abundant vitality and fecundity which every healthy organic species possesses. Hence the occasional destruction of colonies by the parasites does not seriously endanger the life of the host species. If this were the case the hosts would be scarce or would have disappeared long ago. The same considerations probably apply to the human species, for it would seem that only man's world-wide distribution, great fecundity and wonderful adaptability have enabled him to survive all the terrible exploitations to which he has been subjected.

LECTURE VI

THE TERMITES, OR "WHITE ANTS"

THE twenty-one independent social organizations which I have reviewed in the preceding lectures all belong to two natural orders, the Coleoptera and the Hymenoptera, insects so complex and specialized in structure and in their postembryonic development that they are placed by entomologists among the most advanced and recently evolved groups. In order to complete our survey and include the three remaining types of social life, we shall have to turn back to the opposite end of our classification, to the most ancient and primitive orders, the Dermaptera, Embidaria and Isoptera, insects that do not hatch from the egg as specialized and often aberrant maggot-like larvæ, but as forms very much like the adults, except that they lack wings. Owing to their archaic character, it would have been more natural to deal with these three orders in the first lecture, but I have reserved them till the end because one of them, the Isoptera, or termites have a social organization in some respects even more highly developed than that of the Hymenoptera and are best understood by comparison with the ants. The two other orders, the Dermaptera and Embidaria, have a very feeble or rudimental social organization and may, therefore, be treated very briefly. They are interesting mainly as showing that social tendencies must have made their appearance very early in the ancestral history of the insects and as indicating that such tendencies may have been rather prevalent among the ancient Carboniferous orders.[1]

The Dermaptera, or earwigs are peculiar elongate insects, provided with strong forceps at the posterior end of the body, very short, leathery fore wings, or elytra and singular fan-shaped hind wings. Most of the species are tropical and vegetarian, though some are more or less omnivorous or even predatory. The forceps, which differ in the two sexes, are used both in defense and in seizing the prey. We have very few native species of Dermaptera. The common European earwig, *Forficula auricularia,* has been recently introduced into Newport, R. I., where it has become a serious pest in gardens and houses, and another large European form, the maritime earwig, *Anisolabis maritima,* is now common along our coast from Maine to Long Island. You will find it abundant at Revere Beach at or near highwater mark under the driftwood and débris tossed up by the waves. A very small native species, *Labia minor,* is common in manure heaps, mushrooms, etc.[2]

The earwigs may be included among the subsocial insects because the female carefully guards, rearranges and licks her eggs and even remains with the young for a short time after they hatch. The habits of the common European species have been repeatedly described. The German entomologist Frisch, as early as 1730, and the more eminent Swedish entomologist DeGeer in 1773, recorded the fact that the female cares for her eggs and young. The following more detailed observations on the maritime earwig were published by Bennett in 1904: " When about to lay her eggs the female would make a little chamber for herself in the ground about half an inch deep and one, or one and one quarter, inches wide. This was hollowed out beneath a log or some other object that rested on the ground. In making this chamber she carried the earth out in her mouth-parts, as already suggested, a little at a time, just as an ant would do. She never seemed to use her forceps for either digging or carrying the earth. The chamber is made perfectly clean; no sticks or bits of wood or pebbles are allowed by the more careful females to remain inside. Here she deposits her eggs. In the chamber of one wild earwig I counted about ninety eggs, but none of those I have had in captivity laid quite so many at a time, some laying only twenty-five or less. Immediately after the eggs were laid the female picked them up in her mouth-parts, one at a time, and wiped them all over. It looked indeed, as if she rolled them in her mouth. However that may be, when the process is over the eggs are all clean and glossy. Then she places them in a neat pile and stands guard over them. Whether or not it be true with the wild insect, some of the females I have kept for observation have, before their eggs were hatched, moved them all several times from place to place, carrying them one at a time. Some of my earwigs refused to touch food of any kind, so far as I could see, from the time they laid their eggs until the young were hatched, while others would leave their eggs at times to get something to eat. Several times the females, after caring for their eggs a while have eaten them. I have reason to think, however, that in nearly every case the eggs had already spoiled or dried up before this oc-curred. The females continued to guard their young for a few days after they were hatched. When, however, they had once left her to seek food for themselves, they could not safely return lest she would endeavor to eat them. One earwig which I kept in confinement deposited four fairly large batches of eggs in one summer."

The Embidaria (Fig. 93) are a very small order of some

11 genera and 56 described species, confined to the tropics and of an even more archaic habitus than the earwigs.[3] They are rare or local and live in small colonies in peculiar silken webs which they spin in cavities of the soil under stones or on the bark of living trees. The females are al-

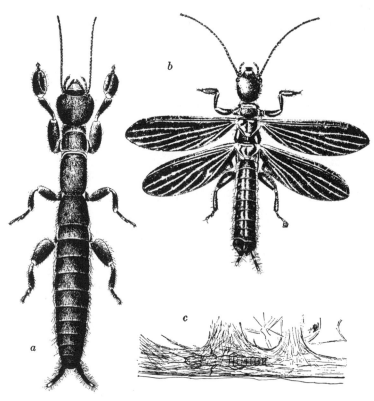

Fig. 93. — a, *Embia major* of India, female; b, male. (After A. D. Imms.) c, *Anisembia texana* of Texas, female in web. (After A. L. Melander.)

ways wingless, but the males in some species are dimorphic, one form having well-developed wings, the other being wingless. They feed on vegetable substances. The webs (Fig. 93c) consist of tubular, branching or anastomosing

galleries and are spun with the fore feet which have peculiar silk-glands in their enlarged metatarsal joints. I have
seen these webs in great numbers on the bark of rubber
trees in the island of Trinidad and of various forest trees
in Guatemala.

The habits of the Embidaria have been studied by
Grassi and Sandias, de Saussure, Melander, Perkins, Friederichs, Kusnezov and Imms, and we possess two excellent
treatises on the morphology and classification of the group
by Verhoeff and Krauss. Imms has recently discovered
under stones in the Himalayas a large species, *Embia major* (Fig. 93*a* and *b*), nearly an inch long, and has published
a full account of its habits. He summarizes his observations on the care of the young as follows: " Maternal care
on behalf of the ova and young larvæ is strongly exhibited
by the females, in very much the same manner as has been
long known to occur among the Dermaptera from the observations of Frisch, DeGeer, Xambeu, Green, and others.
The female *Embia major* shows very marked solicitude for
the welfare of her offspring after her first few eggs have
been deposited. She takes up her position in close proximity to the ova and usually concealing them, so far as
possible, by means of her body. If alarmed and driven
away, she returns sooner or later to take up the same attitude. When the young larvæ are hatched they remain
around the parent female, who conceals them, so far as
she is able, by means of her body, very much after the same
manner as a hen guarding her brood of chickens. A female
and her brood were kept in a small glass trough and observed daily living in intimate association. When separated
from the parent the larvæ were observed the next day to
have regained their former position. As the larvæ approach their second stage in growth, they exhibit a ten-

dency to wander away from the female and construct small tunnels for themselves. They are markedly social, the whole of a brood living together within a complex silken meshwork of tubes." [4]

The resemblances in the habits of the Dermaptera and Embidaria will be sufficiently clear from the foregoing remarks. In both orders there is a feeble tendency to a social organization like that of the dung-beetles considered in my first lecture. It should be noted that the female Dermapteron and Embidarian live for several months after mating, so that in their cases also the first prerequisite for an extension of maternal care of the offspring is realized.

In the Isoptera, or termites, we encounter a social organization of unusual interest, both because it is so elaborate, though exhibited by insects of a very primitive anatomical structure, and because it parallels in so many of its features the social organization of the most highly specialized Hymenoptera. It is as if we had found, when Australia was first explored, the kangaroos and opossums enjoying a social organization like that of man.

The study of the termites began in the eighteenth century with König, who described some of the Indian species in 1779, and Smeathman, who in 1781 published an excellent paper on some West African species. During the nineteenth century the group attracted few investigators. In 1855–60 Hagen, who was later curator of entomology at our Agassiz Museum in Cambridge, published an important monograph, and Lespès in 1856, Fritz Müller in the seventies (1871–1875), Grassi and Sandias and Haviland at the close of the century (1897–98) gave us valuable accounts of the habits and development of the South European, South American and paleotropical species. During

the past twenty years, however, there has been a veritable
"boom" in the investigation of these insects. A long list
of able workers — Wasmann, Desneux, Sjöstedt, Holmgren,
Silvestri, Escherich, Bugnion, von Rosen, Trägårdh, Fey-
taud, Petch, Imms, Nassonov, and Uichanco, in the Old
World, and Andrews, Knower, Heath, Banks, Snyder, Emerson and Miss Thompson in the United States — have
added greatly to our knowledge of the termites. This
speeding up of investigation may be gauged by the number of species brought to light since the time of Smeathman. In the early nineteenth century only a few were
known, in 1858 Hagen cited 60 forms, by 1904 Desneux
was able to list 343 in the "Genera Insectorum," and at
the present time the number of described species must be
over 1,000. No doubt the greater accessibility of tropical
faunas during the past two decades and the increase in the
number of well-trained entomologists adequately accounts
for this extensive and intensive investigation, for the termites are nearly all tropical or subtropical insects, and being soft-bodied and rather unattractive have never ap-
pealed to the amateur who delights in collections of
beautiful pinned specimens. We have only 38 species of
termites in the United States and only two of them reach
the latitude of Boston.[5]

The recent ransacking of the tropics by the specialists
above mentioned has led to a much more satisfactory classification of the Isoptera. According to the latest and best
arrangement, devised by Nils Holmgren, the order may be
naturally divided into four families, the Mastotermitidæ,
Protermitidæ, Mesotermitidæ and Metatermitidæ. All
the species are social, but the four families show a progressive development in the order mentioned and in both structure and habits, from primitive and generalized to more

advanced and specialized forms. I shall refrain from annoying you with taxonomic details, but I must dwell for
a moment on the Mastotermitidæ, which have the same
interest for the termitologist that *Pithecanthropus erectus*
has for the anthropologist. The family Mastotermitidæ was
established by Froggatt for a single species, *Mastotermes
darwiniensis,* from Northern Australia. A study of its
structure shows that it combines characters peculiar to
the cockroaches on the one hand and to the true termites
on the other. Its general habitus is cockroach-like, and it
has distinctly five-jointed tarsi and an anal lobe to the fore
wing, cockroach characters which do not appear in other
termites. At the same time, as Holmgren has shown, this
insect is not so clearly related to the modern Blattidæ, or
cockroaches as it is to their ancestors, the Protoblattidæ,
which became extinct in the Permian. We are therefore
justified in regarding the termites as primitive social cockroaches, which probably came off from the ancestral blat-
toid stem in late Paleozoic or early Mesozoic times. Some
light is also thrown on the significance of Mastotermes by
a study of the fossil termites. Although we have no pale-
ontological records of these insects before the Tertiary,
von Rosen has been able to discover four extinct species
of Mastotermes, three (*bournemouthensis, anglicus* and
batheri) in the Eocene and Upper Oligocene of England
and one (*croaticus*) in the Miocene of Croatia. This indicates that the genus, now confined to Northern Australia
and reduced to a single species, once contained many species and was distributed throughout the continents of the
Old and probably also of the New World. Other termite
genera are abundantly represented in the Baltic Amber
and in the Miocene formations of Colorado and Europe,
and von Rosen has shown that they belong to the two

lower families of the higher termites, the Protermitidæ and Mesotermitidæ, the highest family, the Metatermitidæ, which builds the extraordinary nests in the tropics at the present day, being absent from the early Tertiary formations. This does not mean that the Metatermitidæ did not exist at that time but that the European climate during the Oligocene and Miocene may have been too cool for them. Even to-day they are almost exclusively confined to the warmest portions of the earth. In all probability, therefore, the termites, like the ants, reached their complete structural and social development in the late Cretaceous or early Tertiary and have since undergone very little modification.

It is difficult for the inhabitants of temperate regions to appreciate the great importance of the termites in the tropics where they may, at least locally, equal or even surpass the ants in the number of their colonies and individuals. Even the casual visitor to the tropics will be impressed by the abundance of ants, but he may never see the termites, because the wingless members of their colonies rarely expose their pale, thin-skinned bodies to the light and air, but keep them carefully concealed under the soil, in covered galleries or in the dead and rotting vegetation. He may be astonished to see the dark-colored, winged, sexual forms suddenly appear at certain times in great numbers, but unless he is an entomologist, he will probably regard them as true ants. Closer observation, however, reveals the fact that the activities of the termites are both extremely helpful and extremely injurious to man. In South Africa Drummond long ago found that they perform an important function like that of the earthworms in moist temperate regions, but on a vaster scale. Since they are so very numerous and feed almost exclusively on dead vegetable substances they conspire with the bacteria, high temperature and humidity to accelerate the disintegration of all the lifeless plant matter and to convert it into humus which can be at once utilized by the growing vegetation. The termites are therefore important agents in assisting the growth and renewal of the great

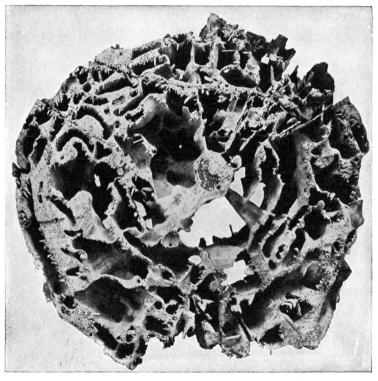

Fig. 94. — Damage by *Calotermes hubbardi* to rafters in an " adobe " building in Arizona. (After Thos. E. Snyder.)

rain-forests of the Amazon, the Congo and the East Indies. But even in the dry savannas of South America and Africa and the open forests of Australia they hasten the dissolution of the dead grasses and other herbaceous plants as well as that of the sparse bushes and trees. On the other hand their insatiable appetite for cellulose and anything containing it makes these insects a terrible menace to all wooden constructions, such as fences, telegraph poles, houses, railway ties, bridges, furniture, ships, fabrics and books (Figs. 94 and 95). They do millions of dollars worth of damage annually. It has even been claimed that their

FIG. 95.—Book from the library at Van Buren, Arkansas, ruined by termites. (After Thos. E. Snyder.)

fondness for literature is in part responsible for the slow cultural growth of many tropical countries. Alexander von Humboldt states that he rarely saw books more than 50 years old in Northern South America. The termites have undoubtedly obliterated so many of the records of human achievement that they must be regarded as the subtlest enemies of the historian and archeologist. The damage is so serious that progressive tropical communities are now beginning to use stone, cement and iron instead of wood in the construction or at least in the foundations of buildings.

Termite, like wasp, bee and ant colonies, show an increase in population as we pass from primitive species, such as the Protermitidæ, to such highly specialized form

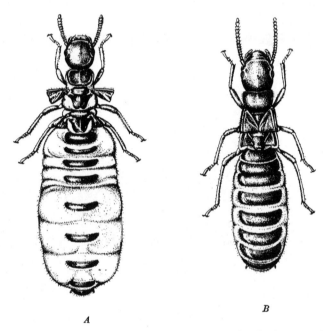

A

B

FIG. 96.—*Reticulitermes flavipes. A.* Old, deälated, physogastric first form queen. *B.* Old, deälated, first form king. (After N. Banks and T. E. Snyder.)

as the Metatermitidæ. Thus the colonies of Calotermes, Archotermopsis and Termopsis consist of only a few dozen or a few individuals, whereas those of the higher termites may comprise hundreds of thousands. Andrews estimated the number of individuals in a moderately large colony of *Eutermes pilifrons* as 631,878. The populations of some African, Indian and Australian Metatermitidæ probably run into the millions.

The polymorphism, or caste development is strikingly like that of the ants, but closer study shows that the resemblances really conceal remarkable and complicated differences. In ants we distinguished only three or four castes: males, queens, workers and soldiers, the last appearing only in certain species. Only one of the castes is

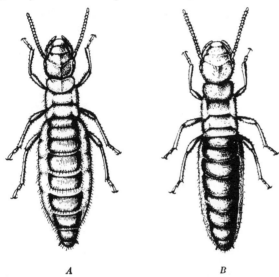

A   *B*

FIG. 97. — *Reticulitermes tibialis.* *A,* incipiently physogastric, apterous queen of the second form; *B,* Apterous king of the second form. (After N. Banks and T. E. Snyder.)

male and is socially unimportant, while the three others are exclusively female. In the termites the sexes are of equal social importance, because each caste comprises individuals of both sexes, though they are almost or quite indistinguishable externally. In the great majority of species there are five castes, three of which are fertile and two sterile. They may be briefly characterized as follows:

1. *First form adults* (Fig. 96), usually called kings and queens. They are very similar, deeply pigmented insects, with large compound eyes, large brain and frontal gland, well-developed reproductive organs and at first with well-developed wings, but these later break off at preformed

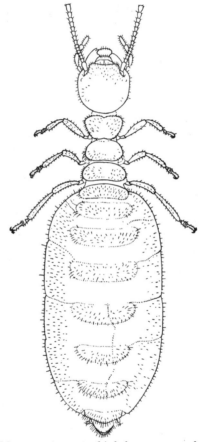

Fig. 98.— Mature, apterous, third form queen of *Reticulitermes flavipes*. (After N. Banks and T. E. Snyder.)

basal sutures and are discarded. Old individuals of this caste therefore can always be recognized by the truncated wing-stubs.

2. *Second form adults* (Fig. 97), sometimes called neo-teinic, complemental, or substitutional kings and queens. Less pigmented, with wingpads, or incipient, nonfunctional wings; brain, compound eyes, frontal gland and reproductive organs somewhat smaller than in the first form.

3. *Third form adults* (Fig. 98), sometimes called ergatoid complemental, or substitutional kings and queens.

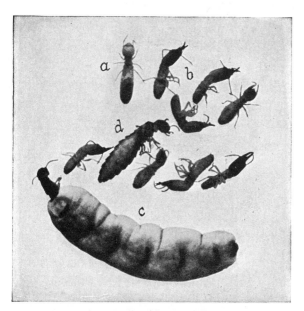

Fig. 99. — *Leucotermes tenuis* of the Bahamas. x2. *a*, worker; *b*, soldier; *c*, physogastric queen; *d*, king.

Scarcely pigmented; entirely wingless. Brain small; eyes and frontal gland vestigial; mature reproductive organs smaller than in the second form.

4. *Workers* (Fig. 99*a*). Wingless, unpigmented; brain small, compound eyes and frontal gland extremely small or absent; reproductive organs embryonic, nonfunctional. Head broader than in the first and second adult forms.

5. *Soldiers* (Fig. 99b, Fig. 103). Wingless; head large and more or less pigmented; brain very small; compound eyes vestigial; frontal gland and in very many species the mandibles and mandibular muscles large (mandibulate soldiers). In a few genera the soldier caste is represented by a different form, usually smaller than the worker, with retort-shaped head, produced anteriorly in the form of a long, tubular snout, with the opening of the large frontal gland at its tip. This form is known as a " nasutus."

The following deviations from this typical series of castes may be noticed. In some Protermitidæ (Calotermes, Archotermopsis, Termopsis) the worker caste is absent and its place is taken by the young soldiers and sexual forms. In another genus, Anoplotermes, the soldier caste is absent though typical workers are present. In some genera (Eutermes, Rhinotermes, Acanthotermes, Capritermes, Termes) there are two or even three different types of soldiers, and in a few genera (Synacanthotermes, Termes) there are two different types of workers. Hence we have the following eight castes, each represented by males and females and therefore making 16 different kinds of individuals: .

1. First form males and females (true kings and queens).
2. Second form males and females.
3. Third form males and females.
4. Large male and female workers.
5. Small male and female workers.
6. Large male and female soldiers.
7. Medium-sized male and female soldiers.
8. Small male and female soldiers.

Probably no single colony of termites ever produces all of these castes, but five or even six of them are frequently represented. Attention should be called to the fact that the soldiers and workers are really larval (nymphal) forms that have been arrested in their development and peculiarly differentiated in certain characters. The second and third adult forms are also arrested, at least so far as the development of their wings is concerned. Owing to the fact that their bodies are still immature or larval, though their reproductive organs may be functional, they have been regarded as " neoteinic."

Although the origin and meaning of the various castes in termites is a matter of considerable interest and has given rise to much discussion, I shall have to treat it very briefly. It was formerly supposed that all termite eggs were alike and therefore produced young larvæ which were at first the same but took on the various caste characters as a result of differences in feeding (trophogeny). Recently, however, Bugnion and especially the late Miss Caroline Thompson of Wellesley have shown that the castes are at least to an important degree determined in the egg. Miss Thompson found that the very young nymphs, or larvæ on hatching may be divided into two series, one with larger, the other with smaller brain, eyes and reproductive organs, and that the former give rise to the reproductive, the latter to the soldier and worker castes (Fig. 100). In all probability even closer study will show that each of the three reproductive and each of the two sterile castes may be distinguished at the time of hatching by certain slight internal characters. The observations of Miss Thompson and Snyder also indicate that each of the adult sexual forms can reproduce itself and the forms below it in rank, but not those above it. Thus the first form adults, or true royalty, can reproduce all the castes, the second form adults can reproduce their own form, the third form, the soldiers and workers, and the third form adults can reproduce only themselves, the soldiers and workers. The soldiers and workers are normally sterile, but sometimes they become fertile and on such occasions probably reproduce only their own castes. It would seem, therefore,

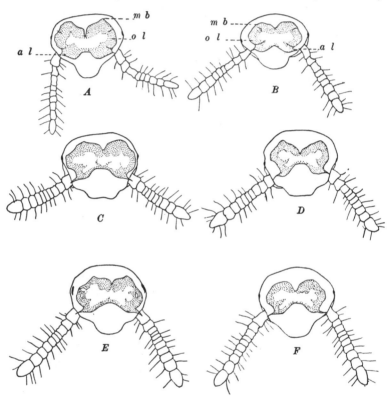

Fig. 100. — Sections of heads of young termites to show relative size of brain and eyes. *A, C* and *E,* just hatched reproductive forms; *B, D, F,* just hatched worker-soldier forms. *mb,* mushroom body of brain; *ol,* optic lobes; *al,* antennary lobes. (After Caroline B. Thompson.)

that the various castes, except the first, are really mutations, partly fertile and partly sterile, and comparable with the series of mutations which certain plants tend to produce in each generation (*Œnothera lamarckiana*). This brings us back to Darwin's discussion of neuter and sterile insects in the eighth chapter of the "Origin of Species," [6] where, referring to the garden stock which produces single and double flowers, he says: "According to M. Verlot, some varieties of the double annual stock from having been long and carefully selected to the right degree, always produce a large proportion of seedlings bearing double and quite sterile flowers; but they likewise yield some single and fertile plants. These latter, by which alone the variety can be propagated, may be compared with the fertile male and female ants, and the double sterile plants with the neuters of the same community. As with the varieties of the stock, so with social insects, selection has been applied to the family, and not to the individual, for the sake of gaining a serviceable end." If further investigation, as now seems likely, extends Miss Thompson's interpretation and shows that each termite hatches from the egg as a recognizable member of a particular caste, we shall have to revise our views on caste differentiation in the ants, because the same explanation may apply to them. It is, in fact, not improbable that the caste of a particular ant is likewise determined in the egg, but special feeding of the larva may be necessary to bring it to maturity. The passively fed ant-larvæ are certainly very different from the active termite nymphs, which soon after hatching are able to run about and seek their own food.

The social functions of the various termite castes correspond to those of the analogous castes among ants. The dark-colored, winged kings and queens, soon after reaching maturity, suddenly leave the nest in a swarm and fly up into the air. This has been called a nuptial flight from analogy with the flight of the male and female ants, but as it is not accompanied by mating, it is more properly

called a flight of dispersion.   The kings and queens after
spending a very short time in the air, alight on the ground,
throw off their wings and associate in couples, the king
running along after the queen.   As soon as they find a
convenient spot each couple digs a small chamber
(Fig. 101).   Both insects work together on its excavation,
and it is not till their habitation is completed that mating
occurs.   The whole termite colony is the offspring of such
a royal pair, just as the ant-colony arises from the single

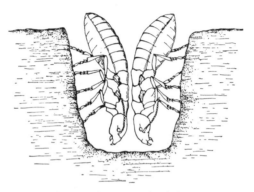

Fig. 101. — Young king and queen of *Hodotermes turkestanicus*
beginning to dig their burrow in the soil after the dispersion flight.
(After G. Jacobson.)

fecundated queen.   In the lower termites the queen undergoes little change in shape since she produces few eggs and
those at rather long intervals, but among the Meso- and
Metatermitidæ, which keep their king and queen in a
special royal chamber in the middle of the termitarium,
the queen's abdomen swells to extraordinary dimensions
as a result of the amount of food she is given and the
enormous growth of her ovaries and fat-body (Fig. 99c,
102, 103).   In some species of Termes she may attain a
length of four inches and 20,000 times the volume of one
of her workers.   Escherich, who has studied the fertile

queens of various Abyssinian and Indian species of Termes, gives a remarkable description of their machine-like — he calls it factory-like — oviposition. Since one of these queens lays an egg every few seconds with clock-like regularity, he estimates that she must lay about 30,000 a day,

FIG. 102. — Three physogastric queens of *Termes*, sp. in the same royal chamber. (From a photograph taken by H. O. Lang at Niangara in the Belgian Congo.)

10 million a year and 100 million during her lifetime of about ten years. Certainly the Termes queen outrivals in fecundity any other terrestrial animal.

The second and third form adults are usually interpreted as substitute royalty, *i.e.*, as forms which may take the places of the true king and queen if they happen to die.

Fritz Müller found in the nest of a South American Eu-termes a true king with a harem of 31 queens of the second form, and Holmgren describes another South American ter-mite, *Armitermes neotenicus,* in the colonies of which he constantly found a true king with a harem of about 100 females of the second form. The Swedish investigator be-lieves that in this species true queens are no longer produced.

The function of the worker termites is like that of the worker ant, mainly concerned with feeding and cleaning the other castes and constructing and repairing the nest. The soldiers of the mandibulate type are evidently defense organisms which use their powerful jaws like the soldiers of ants, but the nasuti, which are usually smaller than the workers and have vestigial mandibles, resort to a dif-ferent method of fighting. Andrews, who studied the be-havior of *Eutermes pilifrons* in Jamaica, says that the nasuti " are potent defenders of the community, at least against other communities. Nasuti were seen to attack alien nasuti and workers by thrusting their snouts against and close to the enemy and ejecting a minute amount of liquid from the tips. This liquid is perfectly clear and colorless. . . . When a coverglass was held over excited soldiers this secretion was collected as if by squirting across space. When drops of it were injected by the soldier onto the heads of other soldiers and workers it seemed to pro-duce a sort of paralysis, which in some cases was connected with the adhesion of the antennæ to the head. . . . The secretion seems to act merely mechanically to stick legs, etc., till the foe was powerless."

The feeding habits are very complicated. Although dead wood and other dead vegetable substances are gnawed off and ingested by the individual termites, they have also de-

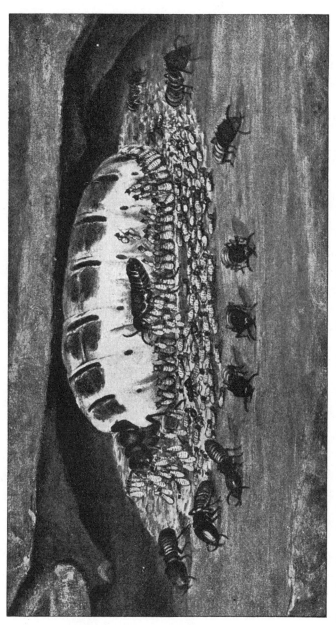

FIG. 103.— Scene in the royal chamber of the African *Termes bellicosus* showing the king, physogastric queen and the attendant soldiers and workers. (From a water-color drawing by G. Kunze. (After Prof. Karl Escherich.)

veloped a very elaborate system of mutual feeding (trophallaxis). They feed one another with saliva, with regurgitated, partially digested food (so-called stomodæal
food) and with food that has passed through the alimentary canal, *i.e.*, with feces (proctodæal food). But even
this does not exhaust the nutritive resources of these extraordinary insects. Holmgren finds that all the castes
produce exudates, that is fatty substances which, arising
from the fat-body and blood-plasma, exude through the
thin chitinous investment of the abdomen on to its surface
and are licked up by other members of the colony. There
seem to be as many different kinds of exudates as there
are castes, and probably the castes are recognized by their
taste.   The physogastric queen produces the most deli-
cious and most copious exudate and is therefore constantly
surrounded and licked by a host of workers (Fig. 103).
According to Escherich, her exudate is so attractive that
the workers in order to make it flow more abundantly will
even tear little strips out of the royal hide. He also states
that many old queens bear the scars of such treatment as
small brown spots or freckles on their milk-white bodies.
We have seen abundant evidence of trophallaxis among the
ants and social wasps, but in none of them is it so elaborately developed as in the termite colonies, the members
of which may be said to be bound together by a circulating
medium of glandular secretions, fatty exudates and partly
and wholly digested food, just as the cells of the body of
a higher animal are bound together as a syntrophic whole
by means of the circulating blood.

Some of the most extraordinary idiosyncrasies of the
termites are exhibited in the construction of their nests,
the termitaria. These bear a certain albeit superficial resemblance to ant nests, but express very clearly the cryp-

tobiotic tendencies of the termites, their great predilection for darkness, moisture and protection from air-currents. Having shut themselves off as much as possible from the outer world and having therefore become almost or quite blind and very thin-skinned, they would be exterminated by predatory insects, birds and mammals and especially by their most inexorable enemies, the ants, if they had not learned to compensate for their feeble and defenseless condition by building unusually strong and resistant nests. There are, to be sure, a few species, like the African *Hodotermes havilandi* and the Indian *Eutermes monoceros*, which are deeply pigmented or have well-developed eyes and forage in the open sunlight, but these are very exceptional.

The nests of termites are of two kinds, the diffuse and the concentrated.[7] The former are not definitely marked off from the environment in the soil or dead wood and consist merely of excavated galleries. Such nests are made by the Protermitidæ and many Mesotermitidæ, *e.g.*, our common *Reticulitermes flavipes*. The concentrated nests are clearly marked off from the environment and are definite, elaborate structures consisting of a royal chamber for the king and queen in the center, surrounded by more or less concentric and inosculating galleries and chambers and an outer covering often of considerable thickness and solidity.

The concentrated nests differ greatly in form in different species. They consist either of earth or woody material or of both, but these substances are subjected to an elaborate preparation. Both the soil and the wood may be swallowed by the workers and after mixture with secretions, either regurgitated or passed through the intestine and used as building material, or particles of soil or wood may be merely bitten off and agglutinated with saliva. On drying the substances employed, especially the saliva-impregnated earth, become almost as hard as cement, so that it is by no means easy to break into some of the larger earthen termitaria. Termites may build either in or on the ground or on the trunks or branches of trees. When in the former situations the nests are typically conical mounds, but their size and shape may vary greatly. In all cases they differ from ant-nests in lacking exposed entrances. When the termites do not pass to or from the nest beneath the surface of the soil, they construct earthen or carton arcades or galleries, so that they can visit their feeding grounds without exposing their bodies to the light and air.

The most conspicuous type of termitarium in the Old World tropics is a rounded boss or cone built up very gradually over the original subterranean nest established by the royal couple and their first brood of workers. Termitaria of this description are often so large and numerous that at a distance they resemble a village of native huts. Closer study of these nests reveals the existence of several different types which may be conveniently grouped as the more or less conical, the columnar or tower-shaped, the wedge-shaped and the mushroom-shaped. In the open savannas of Central Africa a large, broad-based, bluntly pointed, grass-covered termitarium is in some places so common as to make large areas of the soil very uneven (Fig. 104). Even larger, conical masonry nests were described and figured by early explorers like Smeathman. The photographs of a more recent traveler, Mr. H. O. Lang, give a better idea of these structures, which may attain a height of 18 to 20 feet (Fig. 105). In Australia the forms of the earthen nests seem to be even more varied. In New South Wales the structures of *Coptotermes lacteus* first appear above ground as broad, flattened, extremely hard bosses,

Fig. 104.— Mounds of *Termes natalensis* in a grassy plain during the rainy season. Niangara, Belgian Congo. (Photograph by H. O. Lang.)

but they gradually grow up till they form smooth tower-like structures 6 to 10 feet high. The largest of these nests are probably about 10 years old. I infer this from the statements of Machon, who found that similar nests in

FIG. 105. — Pyramidal termitarium of *Termes* sp. at Kwamouth, Belgian Congo. (Photograph by H. O. Lang.)

Paraguay reach a height of 12 feet in 11 years. But there are taller nests in Australia. Those of *Eutermes pyriformis* figured by Saville Kent and Froggatt are veritable skyscrapers 18 to 20 feet high. At Koah, in Northern Queensland, I found another type of columnar nest, also the work of a species of Eutermes but short and bulky, often flattened at the summit and narrowed or constricted at the

Fig. 106. — Mushroom-shaped termitarium of *Eutermes fungifaber*
Medji, Belgian Congo. (Photograph by H. O. Lang.)

base. A transition between this and a more conical type is represented by certain termitaria found by Saville Kent in the Kimberley district of Western Australia. In the northern part of the same continent we find the singular wedge-shaped structures, called compass, or meridional termitaria, because they are constantly oriented with their two long surfaces facing east and west and their narrow

FIG. 107.—Section of same nest as shown in Fig. 106. (Photograph by H. O. Lang.)

ends facing north and south. The peculiar mushroom-shaped termitaria are known only from equatorial Africa. They are only a few inches to a foot in height (Figs. 106 and 107).

The tree nests, which seem to be more abundant in the American than in the Old World tropics, are usually subspherical or ellipsoidal and vary from the size of a football to that of a barrel. They are black or dark brown, consist of digested wood and resemble the carton nests of arboreal ants (Crematogaster, Azteca), except that they have a royal chamber in the center and lack the entrance holes in the covering, or envelope. Certain African and Indian tree termitaria are more cylindrical and resemble stalactites or masses of some viscous substance which has been applied to the tree trunk and has congealed on beginning to drip. In one interesting African species we find on the bark of the tree above a nest of this type a series of chevron-shaped ridges which seem to be constructed by the termites for the purpose of leading the water that flows down the trunk during heavy showers away from the nest (Fig. 108). In British Guiana Mr. Alfred Emerson called my attention to similarly protected nests constructed by an undescribed species of Hamitermes.

Some termites habitually store food substances in the chambers of their nests. Andrews has shown that the common arboreal *Eutermes pilifrons* of the West Indies keeps its food in the form of large solid, lenticular or conchoidal masses in the center of the nest. Some of the African species of Hodotermes (*H. havilandi*) carry in grass, both green and dead, in pieces about two inches long, and the Indian *Eutermes monoceros* collects and probably stores particles of lichens. The Queensland Eutermes that makes the peculiar bulky nests to which I have referred, stores its many chambers with great quantities of cut grass. Silvestri has encountered plant-storing termites in South America.

Of the many analogies between the ants and termites

the most astonishing is that of the fungus-growing habit. Although König and Smeathman, during the latter part of the eighteenth century, independently and almost simultaneously discovered the termite fungus gardens, their true

FIG. 108. — Arboreal nest and tunnels of a termite. Niapu, Belgian Congo. (Photograph by H. O. Lang.)

significance has been appreciated only during the past 30 years, as a result of the investigations of Haviland, Holtermann, Sjöstedt, Petch, Doflein, Escherich and Bugnion.[8]

Fig. 109. — Mushroom gardens of *Acanthotermes militaris* from a nest at Malela, Belgian Congo. The white dots are the food bodies. (Photograph by H. O. Lang.)

While the fungus-growing ants, as stated in a previous lecture, are all confined to a single Myrmicine tribe, the Attiini, and are exclusively American, the fungus-growing termites all belong to a few genera (Termes, Odontotermes, Microtermes, Acanthotermes, Synacanthotermes, Protermes, Sphærotermes) and are confined to the Ethiopian and Indomalayan regions. The fungus-growers, in fact, represent the most highly specialized members of the order Isoptera and are the ones that make the huge nests of which I have cited several examples. In section the nests are seen to contain a number of large, spherical chambers surrounding the royal cell and connected with it and with one another by galleries. In each chamber there are one or more fungus gardens — sponge-like bodies varying in size from that of a walnut to that of a cocoanut, and resembling the gardens of the Attiine ants, but more solidly and more artistically constructed (Fig. 109). They consist of vegetable material which has been collected and comminuted by the workers, passed through their intestines and built up in such a manner as to present the maximum exposure of surface for the growth of the fungi. Petch, who has studied the fungi cultivated by *Odontotermes redemanni* and *obscuriceps* in Ceylon, describes them as growing on the substratum in the form of a mycelium studded with little clusters of swellings like the food-bodies in the gardens of the Attiini. He has succeeded in growing the fungi in the absence of the termites and finds that they belong to at least two species of mushrooms, *Volvaria eurhiza* (Fig. 110) and *Xyglaria nigripes*. According to Bugnion, the mycelium is sown automatically by the worker termites, since they feed on fungus-infected wood and the conidia pass through their bodies without injury. The fungus-gardens are really the nurseries of the termitarium and are full of just-hatched young, which crop the food-bodies like so many little snow-white sheep. Neither the workers nor the soldiers feed on the fungus, but the king

Fig. 110. — Mushroom (*Collybia albuminosa* = *Volvaria* (*Armillaria*) *eurhiza*) growing from an abandoned comb of *Odontotermes* sp. at Coimbatore, India. (Original photograph by W. McRae, reproduced by T. B. Fletcher.)

and queen and other reproductive forms receive the same food as the young.

The nests of some fungus-growing termites are provided with chimney-like structures, communicating with large tubular cavities which have been interpreted by Escherich as a system of ventilating shafts. This interpretation seems to be supported by the existence of somewhat similar arrangements in the nests of the large fungus-growing ants of the genus Atta, and the probability that the successful cultivation of fungi in subterranean chambers depends on a careful regulation of temperature and humidity. Both Petch and Escherich have shown that the diurnal temperature in the termitarium varies only some 9 degrees though the outside temperature may vary as much as 20 degrees or more. On the other hand, the nests of many fungus-growing termites have neither chimneys nor shafts. Trägårdh and Holmgren therefore regard them primarily as passages for the transportation of materials while the nest is being constructed.

Turning now to the association of termites with alien insects we find another striking parallelism with the conditions in the ants described in my last lecture, but again with significant differences. In the tropics termites are very often found nesting in the walls or even in appropriated galleries of nests constructed and inhabited by other species of termites. Holmgren records as many as eight species thus living in the same termitarium. Moreover, many ants, like the species of Carebara, may also live in termitaria. Escherich cites a number of Indian ants as occurring only in such situations, and in Queensland I found fully a dozen species that seemed regularly to inhabit the nests of *Coptotermes lacteus* and allied termites. Certain termites, too, seem to occur only in the nests of certain other termites, *e.g.*, *Anoplotermes fumosus* of Southern Texas and Northern Mexico, according to Snyder, and *Eu-*

termes microsoma, which has similar habits in South America, according to Holmgren. Silvestri found that the South American *Mirotermes fur* always usurps part of the nest and steals the stored food of *Eutermes cyphergaster.* In India Escherich found species of four genera (Leucotermes, Eutermes, Eurytermes and Hamitermes) living in the nests of the fungus-growing species of Odontotermes. With these same termites occur also two species of Microtermes (*globicola* and *obesi*) which form small fungus gardens in their own galleries, though they steal both the fungus and the materials for the substratum from their larger neighbors. In all these cases, whether of termites associated with other termites or of termites associated with ants, we are dealing only with compound nests. The various species always occupy separate galleries and are usually hostile when brought together. No instances of mixed colonies, analogous to those occurring among the ants, have as yet been detected among termites.

More interesting are the relations of the termites to the various insects that live in their nests, the termitophiles, of which several hundred species have been described.[9] They belong to the most diverse orders and families and their association with the termites has evidently been brought about by conditions very similar to those that have induced the myrmecophiles to live with the ants. The termitophiles may also be classified in the same manner, as predators, indifferently tolerated guests (synœketes), true guests (symphiles) and parasites. Some of the forms seem to have undergone little or no modification as a result of their association with the termites, but many of the true guests and a few of the predators have acquired peculiar characters, the most characteristic of which is physogastry, or excessive enlargement of the abdomen. This may be due to a great increase in the volume of the reproductive organs or the alimentary canal but in its most typical form it is brought about by hypertrophy of the fat-body. In correlation with the abdominal enlargement there is a decrease in the size of the head and thorax and a reduction or loss of the eyes and wings. In certain species the body may be furnished with segmental, finger-shaped exudate organs, which are occasionally developed even in physogastric forms. I select for illustration a few of the more striking symphiles and predators:

1. A number of small two-winged flies belonging to the family Phoridæ and the genera Termitoxenia (Fig. 111 A), Termitomyia and Ptochomyia (Fig. 111 B), which live in the nests of African and Indian termites, show an excessive physogastry, with the accompanying diminution of the head, thorax, eyes and wings. The wings, in fact, are reduced to small strap- or hook-shaped vestiges. Wasmann, who first described the species of Termitoxenia and Termitomyia, is of the opinion that they are hermaphroditic and viviparous, but Silvestri has recently shown that this is certainly not the case with the closely allied Ptochomyia. He has also discovered in the nest of a South American termite (*Anoplotermes reconditus*) another physogastric fly, *Termitomastus leptoproctus* (Fig. 111 C), belonging to an entirely different family (Termitomastidæ). It shows a similar physogastry and reduction in the size of the head and thorax, but the eyes and wings have suffered less diminution.

2. Physogastry of various degrees is also exhibited by many beetles of the family Staphylinidæ and subfamily Aleocharinæ, which are fed and licked by the termites. The most extraordinary forms belong to the genera Corotoca and Spirachtha, established many years ago by Schiödte

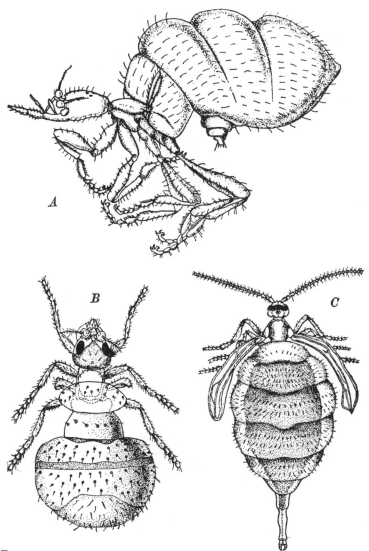

Fig. 111.—Physogastric termitophilus flies. *A. Termitoxenia heimi*, a Phorid from the nest of the Indian *Termes obesus*. (After E. Wasmann.) *B. Ptochomyia afra*, a Phorid from the nests of the West African *Ancistrotermes crucifer*. (After F. Silvestri.) *C. Termitomastus leptoproctus*, a Nematoceran fly from the nests of *Anoplotermes reconditus* in Southern Brazil and Argentina. (After F. Silvestri.)

for three South American termitophiles. In all of them the abdomen is enlarged to form a huge, subspherical mass, which is turned upward and forward over the head and thorax and in Spirachtha bears three pairs of peculiar finger-shaped exudatoria (Fig. 112). When the insect is

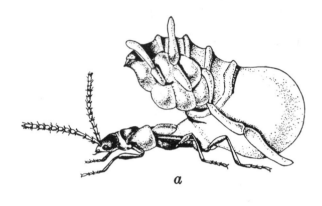

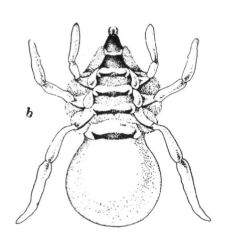

Fig. 112.—*Spirachtha eurymedusa*, a termitophilous Staphylinid beetle from South America; *a*, lateral view of the whole insect; *b*, dorsal view of abdomen, showing the three pairs of appendage-like exudatoria. (After J. C. Schiödte.)

viewed from above (Fig. 112b) only the inverted ventral surface of the abdomen and its appendages are visible. During the summer of 1920 Mr. Alfred Emerson discovered in British Guiana a still more remarkable species of Spirachtha (Fig. 113b and c), which has the two anterior pairs of exudatoria greatly swollen at their tips and the posterior pair elongated and lyriform. The strangest fact

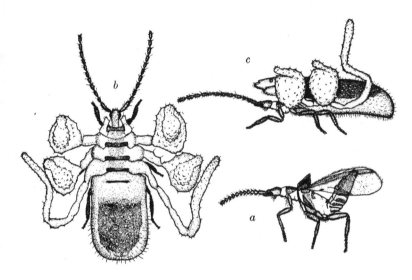

Fig. 113. — Termitophilous Staphylinid beetles from the nests of *Nasutitermes* (*Constrictotermes*) *cavifrons* in British Guiana. *a, Spirachtha schioedtei* Mann (MS.), recently emerged beetle in profile; *b, Spirachtha mirabilis* Mann (MS.), physogastric form with three pairs of abdominal exudatoria homologus with those of Fig. 112; *c,* same in profile. Probably *a* is the young adult of *b;* though the antennæ differ. (From drawings by Alfred Emerson.)

about this beetle seems to be that when it emerges from the pupa it looks like an ordinary Staphylinid (Fig. 113a) but gradually acquires both physogastry and exudatoria during its imaginal life among the termites. Probably this is also the case with other physogastric termitophiles.

3. The larvæ of certain African and Indian beetles of the Carabid genera Glyptus and Orthogonius, which devour the termites among which they live, show a distinct but less pronounced enlargement of the abdomen.

4. The larvæ of several beetles (Staphylinidæ, Cantharidæ), flies (Anthomyidæ) and moths (Tineidæ), which can scarcely be called physogastric, though the abdomen is well-developed, have, like the adult Spirachtha above described, paired finger-shaped and sometimes jointed exudatoria on the abdominal or even on the more anterior segments. Good examples are the larvæ of the Staphylinids, *Paracorotoca akermanni*, described by Warren, and *Œdoprosoma mirandum* Silvestri, the larva of the Anthomyid *Epiplastrocerus mirandus* Silvestri and of the Tineid moths *Plastopolypus integer* and *divisus* described by Trägårdh and Silvestri.

You will observe that the physogastric termitophiles just described resemble their termite hosts and especially the old queens. They also resemble certain ant-larvæ and the queens of Doryline ants. It would seem therefore that all these convergent cases of abdominal enlargement and accumulation of fat, diminution of head and thorax, blindness and aptery must be the results of living in the same peculiar environment. And we should not be far wrong in maintaining that the factors responsible for such modifications are confinement within the narrow galleries and chambers, a very limited supply of oxygen, absence of light and an abundance of carbohydrate food. We have long known that lack of exercise and plenty of food predispose both man and his domestic animals to obesity. That darkness also favors deposition of fat in domestic animals, is shown by the experiments of Ottramare,[10] and we have all heard of the fattening of the geese in the cellars of Strassburg and of the fat-bird (Steatornis) that lives in the caves of Trinidad. Absence of oxygen, moreover, as Dewitz, Bohn and Drzwina have demonstrated, inhibits the development of wings in insects.[11] Andrews says that " the respiratory needs of termites must be slight, since the estimated amount of air in a nest (of the Jamaican *Eutermes pilifrons*) weighing 40 pounds or occupying 4 cubic feet was only 9 volumes for each volume of termite." Whether the partial blindness of many physogastric forms is due to absence of oxygen, as Loeb has suggested in the case of the cave animals,[12] or to the absence of light, as has been universally assumed, can not be decided because both conditions obtain in the nests of termites. Of course, the diminution of the eyes and wings may partially account for the reduction in the size of the head and thorax.

Now the obviously degenerative or pathological phenomenon of fatty physogastry and the development of exudatoria enable the termitophiles to become true guests, *i.e.*, they can, like the termites themselves, produce exudates which are eagerly devoured by their hosts and in return either receive regurgitated food or manage to prey on the defenseless brood. Wasmann believes that the physogastry and exudatoria are produced through " amical selection " by the termites, much as fat breeds of pigs and cattle are produced by man's selective activities, but such an hypothesis seems to me even less acceptable than the very similar hypothesis which he advanced to account for the symphilic myrmecophiles.

Of more importance in the lives of the termites than their physogastric guests are the numerous intestinal Infusoria which have been studied by more than a score of investigators since they were first described by Lespès in 1856 and Leidy in 1877.[13] These Infusoria occur as a rule only in the soldier and worker. They have been found in many species of termites in all parts of their range and have been variously interpreted as parasites, commensals and symbionts. Imms, who has published the most recent account of their behavior, regards them as true symbionts, which break down the particles of ingested wood and render them more easily assimilable by the termites. Animals as high in the scale as insects must find it difficult or impossible to digest crude cellulose. It is therefore interesting to observe, Imms remarks, that the symbiosis between the intestinal Protozoa and the termites is paralleled by the occurrence of numerous genera of Infusoria in the stomachs of ruminants, notably of the ox, sheep, goat, camel and reindeer. " It is believed that, by means of their action upon the vegetable matter consumed by the Ruminants, these Infusoria help to render it capable of being digested by the latter. Furthermore, Infusoria are absent from the stomachs of the young Ruminants prior to being weaned from their parents (*vide* Neveu-Lemaire, 1912, p. 446). According to Certes (1889), glycogen is present in the protoplasm of the Infusoria, and the latter perform a special rôle in the digestive process of the Ruminants. Gruby and Delafond (1843) maintain that the protoplasm of the Infusoria is itself digested, and thereby contributes towards the nutrition of the host Ruminant. Similarly, the Infusoria inhabiting the large intestine of the Equidæ are possibly symbiotic in their relations with their host."

In conclusion I shall have time to dwell on only a few of the many considerations suggested by the singular parallelisms or convergencies between the termites and the ants, such as the development in both of wingless worker and soldier castes, similar nesting and fungus-growing habits, trophallaxis, relations to guests, etc. The duplication of these phenomena in groups so wide asunder that they are placed by the systematists at the opposite poles of our classification of insects, may be of some interest to the anthropologist, because the study of human cultures reveals the same or very similar institutions and linguistic peculiarities in geographically widely separated peoples. Some anthropologists attribute such similarities to community of origin, while others insist that they are often inventions of independent origin and development. When we reflect that ants and termites have been able, through slow physiological and instinctive processes, independently to evolve such strikingly analogous peculiarities as those I have described, we can scarcely doubt that different human communities, belonging to the same species and endowed with some intelligence, may frequently have hit upon the same inventions.

A less obvious consideration is suggested by the investigations of Holmgren, who finds that as we advance in our study of the termites from the lowest family, the Protermitidæ, to the highest, the Metatermitidæ, we notice a distinct physical, or morphological deterioration in the species and a concomitant improvement of the nervous system. These changes he attributes to advancing social organization. A study of the ants as a group certainly reveals a similar but much feebler tendency of the same kind as we pass from the more primitive Ponerinæ to the highly specialized Formicinæ and Dolichoderinæ. And it is interesting to note that human society appears to be tending in the same direction. We are becoming accustomed to the thought that our remote posterity will be toothless, hairless and without olfactory organs, toes and possibly other appurtenances, but we console ourselves with the hope that they will have more and better grey matter and a better social behavior than ourselves. We must look forward, however, not only to physical losses in our descendants but to the loss of many of our institutions and customs, for progressive evolution involves not only an acquisition but also a loss of characters. As De Moor, Massart and Vandervelde have pointed out,[14] "all progress must necessarily be attended by degeneration." This explains why the older generation is always scandalized by the young and never remembers that it scandalized its parents.

In termites the amount of degeneration accompanying social evolution is, as stated, much greater than in the ants, and this degeneration seems to have been brought about very largely by an increasing need for protection. With greater elaboration and solidity of nest architecture the termite colonies shut themselves off more and more from the outside world, and all the castes, except the winged males and females, lost their eyes and the tough consistency of their integument. They thus came to resemble the mollusks, crustaceans and certain fishes and reptiles which have withdrawn within a heavy protective armor, and have given up participating in the free competitive and coöperative life of their environment. The ants, with the exception of certain subterranean species, have not been inveigled into adopting this passive and timorous mode of life. Hence it is conceivable that some of the more plastic ant-genera may be or may come to be centers of progressive evolution, but there is every reason to believe that the termites have already reached the end of their course. Until recently we might have been tempted to say the same of certain human races like the Chinese and Koreans, who have shut themselves off for long periods from intercourse with other races, but fortunately the human species is still so young that even centuries of such behavior have failed to leave their imprint on the physical organization, and as long as the more resilient mental and social activities alone are affected, there is always hope of a return to the more open coöperative and competitive life which constitutes the basis of progressive evolution.

DOCUMENTARY APPENDIX

LECTURE I

1. **Kropotkin, A.,** Mutual Aid, a Factor of Evolution. New York, A. A. Knopf 1920 (first ed. 1902). Among those who have advanced a similar conception of the importance of mutual aid, sympathy or altruism, Kropotkin cites the following: **Kessler,** On the Law of Mutual Aid. Lecture to Russian Congress of Naturalists, Jan. 1880; **Lanessan,** La Lutte pour l'Existence et l'association pour la lutte, 1881; **Brandt, A.,** Vergesellschaftung und gegenseitiger Beistand bei Thieren, Virchow's Samml. wiss. Vorträge n.f. Heft 279, 48 pp., 1897. Similar ideas appear, however, in much earlier works, e.g. in **Adam Smith's** Theory of Moral Sentiments, 1759, **Buckle's** History of Civilization, **Lange's** Geschichte des Materialismus and **Vaihinger's** Die Philosophie des Als-Ob, 7th and 8th ed. Leipzig, Felix Meiner, 1922, p. 344 *et seq.* More recently the importance of coöperation in evolution has been stressed by **Kammerer,** Allgemeine Symbiose und Kampf ums Dasein als gleichberechtigte Triebkräfte der Evolution. Arch. Rass. Gesell. Biol. 6, 1909, p. 585–608; **Delage, Y.,** and **Goldsmith, M.,** Les Théories de l'Evolution. Paris, Flammarion, 1914; **Patten, W.,** Coöperation as a Factor in Evolution, Proc. Amer. Phil. Soc. 55, 1916, p. 503–532; The Grand Strategy of Evolution, Boston, R. G. Badger, 1920, and **Reinheimer, H.,** Symbiosis, a Socio-Physiological Study of Evolution, Headly Bros., London 1920, his Evolution by Coöperation which I have not seen, and the anonymous " Glass of Fashion," 1921, p. 145.

2. For accounts of the general phenomena of symbiosis the reader may be referred to the following works: **Van Beneden, J. P.,** Les Commensaux et les Parasites dans le règne animal, 2nd ed. 1878 (Animal Parasites and Messmates. Internat. Sc. Series, 1876); **Hertwig, O.,** Die Symbiose im Tierreich, Jena, 1883; **Laloy, L.,** Parasitisme et Mutualisme dans la Nature, Paris, Alcan, 1906; **Elenkine,** La Symbiose, équilibre instable entre les organismes associés. Trav. Soc. Imp. Nat. St. Pétersbourg 37, 1906 (Russian); **Portier, P.,** Les Symbiotes. Paris, Masson, 1918; **Lumière, A.,** Le Mythe des Symbiotes. Paris, Masson, 1919; **Famintzin, A.,** Die Symbiose als Mittel der Synthese von Organismen. Biol. Centralbl. 27, 1907; **Mereschovsky,** Theorie der zwei Plasmaarten als Grundlage der Symbiogenesis, einer neuen Lehre

285

von der Entstehung der Organismen. Biol. Centralbl. 1910; **Caullery, M.**, Le Parasitisme et la Symbiose, Paris, Doin, 1922.

The following list of works covers a large number of very interesting cases of symbiosis between insects and fungi (excluding the social insects), flatworms and algæ, mitochondria, lichens, orchids and fungi, etc.: **Ames, O.**, Notes on New England Orchids II, The Mycorrhiza of *Goodyeara pubescens*, Rhodora 24, 1922, p. 37–46, 2 pls.; **Becher, E.**, Die fremddienstliche Zweckmässigkeit der Pflanzengallen. Leipzig, Veit u. Co., 1917; **Breest, F.**, Zur Kenntniss der Symbiontenübertragung bei viviparen Cocciden und bei Psylliden. Arch. Protistenk. 24, 1914, p. 263–276, 2 pls.; **Bernard, N.**, L'évolution dans la Symbiose, Les Orchidées et leurs champignons commensaux. Ann. Sc. Nat. Bot. (9) 9, 1909; **Berlese, A.**, Sopra una nuova specie mucidinea parassita del *Ceroplastes rusci*. Redia 3, 1906, p. 8–15; **Brues, C. T.**, and **Glaser, R. W.**, A Symbiotic Fungus occurring in the Fat-body of *Pulvinaria innumerabilis* Rath. Biol. Bull. 40, 1921, p. 299–324, 3 pls., 2 text-figs.; **Buchner, P.**, Tier und Pflanze in intrazellulärer Symbiose. Berlin Gebr. Bornträger, 1921; **Burgeff, H.**, Die Wurzelpilze der Orchideen. Jena, 1909; **Glaser, R. W.**, Biological Studies on Intracellular Bacteria, No. 1. Biol. Bull. 39, 1920, p. 113–145; **Escherich, K.**, Ueber das regelmässige Vorkommen von Sprosspilzen in dem Darmepithel eines Käfers. Biol. Centralbl. 20, 1900; **Fränkel, H.**, Die Symbionten der Blattiden im Fettgewebe und Ei, insb. von *Periplaneta, orientalis*. Dissert. München, 1918; **Frank B.**, Ueber die auf Wurzelsymbiose beruhende Ernährung gewisser Bäume durch unterirdische Pilze. Ber. deutsch. Bot. Gesell. 3, 1885; **Glasgow, H.**, The Gastric Cæca and the Cæcal Bacteria of the Heteroptera. Biol. Bull. 26, 1914, p. 101–151, 8 pls.; **Guillermond, A.**, Mitochondries et Symbiotes. C. R. Soc. Biol. Paris, 82, 1918; **Keeble, F.**, Plant-Animals, a Study in Symbiosis. Cambridge Univ. Press, 1910; **Mercier, L.**, Recherches sur les bactérioides des Blattides. Arch. Protistenk. 9, 1907, p. 346–358, 2 pls.; **Peklo, J.**, Ueber symbiontische Bakterien der Aphiden (Vorl. Mitth.). Ber. deutsch. Bot. Gesell. 30, 1912; **Pierantoni, U.**, L'origine di alcuni organi *d'Icerya purchasi* e la simbiosi ereditaria. Boll. Soc. Nat. Napoli 23, 1909, p. 147–150; Origine e struttura del corpo ovale del *Dactylopius citri*, etc. *ibid.* 24, 1910, p. 1–4; Struttura ed evoluzione dell' organo simbiotico di *Pseudococcus citri* Risso, etc. Arch. Protistenk. 31, 1913, p. 300–316, 3 pls.; La luce degli insetti luminosi e la simbiosi ereditaria. R. C. Acad. Sc. Fis. Math. Napoli 1, 1914; **Putnam, J. D.**, Biological and Other Notes on Coccidæ. Proc. Davenport Acad. Sc. 2, 1880, p. 293–347, 1 pl.; **Rees, M.**, Ueber die Entstehung der Flechte *Collema glaucescens* Hoffm. durch Aussaat der

Sporen desselben auf Nostoc lichenoides. Monatsb. K. Akad. Wiss. Berlin, 1871; **Regaud, C.,** Mitochondries et Symbiotes. C. R. Soc. Biol. 82, 1918; **Roubaud, E.,** Les Particularités de la Nutrition et la vie Symbotique chez les Mouches Tsétsés. Ann. Inst. Pasteur 33, 1919, p. 1–48, 17 figs.; **Shinji, G. O.,** Embryology of Coccids, with Special Reference to the Ovary, Origin and Differentiation of the Germ Cells, Germ Layers, Rudiments of the Mid-gut and the Intracellular Symbiotic Organism. Journ. Morph. 33, 1919, p. 73–167, 20 pls.; **Schwendener, S.,** Ueber die wahre Natur der Flechten. Verh. Schweiz. Naturf. Gesell. 51, 1867; **Sikora, H.,** Vorläufige Mitteilung über Mycetome bei Pediculiden. Biol. Centralbl. 39, 1919; **Sulc, K.,** *Kermincola kermesina* und Physokermina, neue Mikroendosymbiotiker der Cocciden. Sitzb. boehm. Gesell. Wiss. Prag. 1906; Pseudovitellus und ähnliche Gewebe der Homopteren sind Wohnstätte symbiotischer Saccharomyceten. *ibid.* 1910; **Tobler,** Das physikalische Gleichgewicht von Pilz und Alge in den Flechten. Ber. Deutsch. Bot. Gesell. 27, 1909; **Trojan, E.,** Bakteroiden, Mitochondrien und Chromidien. Ein Beitrag zur Entstehung des Bindegewebes. Arch. mikr. Anat. 93, 1919; **Wallin, J. E.,** On the Nature of Mitochondria. Amer. Journ. Anat. 30, 1922, p. 203–229, 1 pl.; p. 451–470, 2 pls.; **Zirpolo, G.,** I batteri fosforescente e la recente recerche sulla biofotogenese. Natura 10, 1919.

3. **Carpenter, Edw.,** Civilisation, its Cause and Cure and Other Essays (2nd edit.). New York, Scribner, 1921.

4. **Wheeler, W. M.,** The Ants of the Baltic Amber, Schrift. Physik.-œkonom. Gesell. Königsberg 55, 1914, p. 1–142, 66 figs.

5. For descriptions of these fossil ants see: **Cockerell, T. D. A.,** A New Fossil Ant. Ent. News 1906, p. 27, 28; British Fossil Insects, Proc. U. S. Nat. Mus. 49, 1915, p. 469–499, 6 pls.; Fossil Arthropods in the British Museum I. Ann. Mag. Nat. Hist. (9) 5, 1920, p. 273–279, 3 figs.; Some Eocene Ants from Colorado and Wyoming. Proc. U. S. Nat. Mus. 59, 1921, p. 29–39, 1 pl., 9 figs.; **Donisthorpe, H.,** British Oligocene Ants. Ann. Mag. Nat. Hist. (9) 6, 1920, p. 81–94, 1 pl.

6. **Barrell, J.,** Rhythms and the Measurements of Geologic Time. Bull. Geol. Soc. Amer. 28, 1917, p. 745–904, 4 pls. See also: **Blackwelder, E.,** The Trend of Earth History. Science, N. S. 55, 1922, p. 83–90, and **Huntington, E.,** and **Visher, S. S.,** Climatic Changes. Yale Univ. Press, 1922. **Chamberlain, T. C., Clarke, J. M., Brown, E. W.** and **Duane, W.,** Symposium on the Age of the Earth. Proc. Amer. Phil. Soc. 61, 1922, p. 247–288.

7. **Jones, F. Wood,** Arboreal Man. 2nd Impr., London, E. Arnold, 1918.

8. For an account of three other groups of insects which should, perhaps, have been considered in this and the following lectures see pp. 290, and 293.

9. **Fiske, John,** The Destiny of Man, viewed in the Light of his Origin. Boston, Houghton Mifflin & Co., 1886, Chapter VI.

10. For a brief but valuable discussion of the appetites see **Craig, W.,** Appetites and Aversions as Constituents of Instincts, Biol. Bull. 34, 1918, p. 91–107.

11. **Berman, L.,** The Glands Regulating Personality. New York, Macmillan, 1921.

12. For this and other cases of apparent insensibility to pain in insects see **Forel, A.,** The Senses of Insects (trans. by M. Yearsley). London, Methuen & Co., 1908, p. 114.

13. **Wallas, Graham,** The Great Society, a Psychological Analysis. New York, Macmillan, 1914

14. See **Fabre, J. H.,** Souvenirs Entomologiques. Vol. I, 1879, Chapters I and II; Vol. V, Chap. I–XII; Vol. VI, Chap. I, II, V and VI and Vol. X, Chap. I–IV. Part of these are reprinted in Fabre, La Vie des Insectes. Paris, Delagrave, Chapters I–X, also translated by A. T. de Mattos in the volume entitled " The Sacred Beetle and Others." New York, Dodd, Mead & Co., 1918. See also Fabre's Book of Insects, retold from the preceding by Mrs. R. Stowell, with illustrations by E. J. Detmold. New York, Dodd, Mead & Co., 1921. A general account of the habits of the various Scarabæidæ studied by Fabre is also given by **Reuter, O. M.,** Lebensgewohnheiten und Instinkte der Insekten bis zum Erwachen der sozialen Instinkte. Berlin; Friedländer, 1913. For an excellent monographic study of the habits and geographical distribution of the dung-beetles see **Kolbe, H. J.,** Ueber die Lebenweise und die geographische Verbreitung der coprophagen Lamellicornier. Zool. Jahrb. suppl. 8, 1905, p. 475–594, 3 pls.

15. **Ohaus, F.,** Bericht über eine entomologische Reise nach Zentralbrasilien. Stettin. Ent. Zeitg. 60, 1899, pp. 204–245; 1900, p. 164–191; 193–274.

16. Descriptions and figures of the stridulatory organs of the larval Passalus will also be found in the following works: **Schiödte, J. C.,** De Metamorphosi Eleutheratorum Observationes: Bidrag til Insekternes Udviklingshistorie. Copenhagen, Thiele, 1862–1873; **Sharp, D.,** Insects. Vol. 2 (in Cambridge Natural History). London, Macmillan 1899; **Wheeler, W. M.** and **Bailey, I. W.,** The Feeding Habits of Pseudomyrmine and Other Ants. Trans. Amer. Phil. Soc. N. S. 22,

1920, p. 235–279, 5 pls., 6 text-figs. The stridulatory organ of the adult Passalus is described and figured by **Babb, G. F.**, On the Stridulation of *Passalus cornutus* Fabr. Ent. News 12, 1901, p. 279–281, 1 pl.

17. **Kirby, W.**, Fauna Boreali-Americana, 1837, p. 188 *nota*. *Phrenapates bennetti* is figured in Griffith and Pigeon's Cuvier's Animal kingdom. Class Insect., Vol. 2, 1832, Pl. 50, Fig. 1, and Pl. 69, Fig. 1.

18. **Champion, G. C.**, Coleoptera, Vol. IV in Biologia Centrali-Americana 1884, p. 139.

19. **Ohaus, F.**, Bericht über eine entomologische Studienreise in Südamerika. Stettin. Ent. Zeitg. 70, 1909, pp. 3–139.

20. See for a fuller account of these insects: **Wheeler, W. M.**, A Study of Some Social Beetles in British Guiana and of their Relations to the Ant-Plant Tachigalia. Zoologica 3, 1921, p. 35–126, 5 pls., 12 text-figs.

21. **Schwarz, E. A.**, and **Barber, H. S.**, Descriptions of New Species of Coleoptera, Zoologica 3, 1921, p. 189–193, 1 pl. The larvæ are described by **Böving, A. G.**, The Larvæ and Pupæ of the Social Beetles *Coccidotrophus socialis* (Schwarz and Barber) and *Eunausibius wheeleri* (Schwarz and Barber), with Remarks on the Taxonomy of the Family Cucujidæ. Zoologica 3, 1921, p. 197–212, 4 pls.

22. This method of constructing a cocoon is most unusual among insects. Recently R. E. Snodgrass has given a very interesting account, with beautiful figures, of a somewhat similar, but much more elaborate method of cocooning, by a small caterpillar (*Bucculatrix pomifoliella*) which feeds on the leaves of the apple (The Resplendent Shield-bearer and the Ribbed-Cocoon-Maker, two Insect Inhabitants of the Orchard. Smithson. Rep. (1920) 1922, p. 485–510, 3 pls., 15 text-figs.). In this case the cocoon is made of silk.

23. For an account of the classification of the Ipidæ see **Hagedorn, M.**, in Wytsman's Genera Insectorum, 1910, 178 pp., 14 pls.

24. The following papers contain the more important observations on the habits of the fungus-growing Ipidæ; **Eichoff, W.**, Die europäischen Borkenkäfer. Berlin, J. Springer, 1881, p. 280–281; **Hubbard, H. G.**, Ambrosia Beetles. Yearb. U. S. Dep. Agric. 1906, p. 421–430, 7 figs.; The Ambrosia Beetles of the United States. Bull. 7 (N. S.) Div. Ent. U. S. Dep. Agric. 1897, p. 9–30, 34 figs.; **Hopkins, A. D.**, On the History and Habits of the Wood Engraver Ambrosia Beetle *Xyleborus xylographus* (Say), *Xyleborus saxeseni* (Ratz), with brief Descriptions of Different Stages. Canad. Ent. 30, 1898, p. 21–29, 2 pls.; **Hagedorn, H.**, Pilzzüchtende Borkenkäfer, Naturwiss. Rund-

schau N. F. 6, 1907, p. 289–293, 4 figs.; **Beauverie, J.**, Les Champignons dits Ambrosia. Ann. Soc. Nat. Bot. (9) 2, 1909, p. 31–73; **Neger, F. W.**, Die Pilzkulturen der Nutzholzborkenkäfer. Centralbl. Bakter. (2) 20, 1908, p. 279–282; Ambrosiapilze, Ber. Deutsch. Bot. Ges. 26a–29: 1. Mitteilung 1908, p. 735–754; 2. Mitteil. 1909, p. 372–389; 3. Mitteil. 1910, p. 455–480; 4. Mitteil. 1911, p. 50–58; Zur Uebertragung des Ambrosiapilzes von *Xyleborus dispar* F. Naturw. Zeitschr. Forst u. Landwiss. 9, 1911, p. 223–225; **Schneider-Orelli, F.**, Die Uebertragung und Keimung des Ambrosia-Pilzes von *Xyleborus* (*Anisandrus*) *dispar* F. Naturw. Zeitschr. Forst. Landwiss. 9, 1911, p. 186–192, 223–225; Ueber die Symbiose eines einheimischen pilzzüchtenden Borkenkäfers (*Xyleborus dispar* F.) mit seinem Nährpilz. Verh. Schweiz. Naturf. Gesell. 94, 1911, p. 279–280; Der ungleiche Borkenkäfer (*Xyleborus dispar* F.) an Obstbäumen und sein Nährpilz. Naturwirtsch. Jahrb. Schweiz, 1912, p. 326–334; Untersuchungen über den pilzzüchtenden Obstbaumborkenkäfer *Xyleborus* (*Anisandrus*) *dispar* und seinen Nährpilz. Centralbl. Bakter. Paras. Infektionskrank. 38, 1913, p. 25–110, 3 pls., 7 text-figs.

25. The classification of the Platypodidæ is summarized in **Strohmeyer, H.**, Family Platypodidæ in Wytsman's Genera Insectorum 163rd fasc. 1914, 55 pp., 11 pls., 1 map.

26. For observations on the habits of Platypodidæ see the papers of Hubbard, cited above and the following: **Strohmeyer, H.**, Beiträge zur Kenntnis der Biologie von *Platypus* var. ? *cylindriformis* Reitt. Ent. Blätter, 3, 1907, p. 65; **Swaine, J. M.**, A New Species of Platypus from British Columbia. Canad. Ent. 1916, p. 97–100, 2 pls.

27. **Strohmeyer, H.**, Die biologische Bedeutung sekundärer Geschlechtscharaktere am Kopfe weiblicher Platypodiden. Ent. Blätter 7, 1911, p. 103.

28. **Schneider–Orelli**, Untersuchungen etc. (*loco citato*).

29. **Handlirsch, A.**, Die Fossilen Insekten und die Phylogenie der rezenten Formen. Text and atlas. Leipzig, Engelmann, 1908.

LECTURE II

1. The Australian saw-flies of the genus Perga may also be included among the subsocial insects. They were first observed by **R. H. Lewis** (Case of Maternal Attendance on the Larva by an Insect of the Tribe Terebrantia, belonging to the genus Perga, observed at Hobarton, Tasmania. Trans. Ent. Soc. London 1, 1836, p. 232–234, and more recently **W. W. Froggatt** has published the following papers on several species: Notes on the Life-history of Certain Saw-flies (Genus Perga)

with Description of a New Species. Proc. Linn. Soc. New South Wales 5, 1891, p. 283–288; The Pear and Cherry Slug (*Eriocampa limacina* Retz), generally known as *Selandria cerasi*, with Notes on Australian Saw-flies. Agric. Gazette N. S. Wales, 1901, p. 1–11, 4 pls.; Notes on Australian Saw-flies (Tenthredinidæ). Proc. Linn. Soc. N. S. Wales 43, 1918, pp. 668–672. Concerning the species observed by Lewis (*Perga lewisi*), Froggatt says (1901) that it is the commonest saw-fly about Sydney on bloodwood (*Eucalyptus corymbosa*) and that the " female makes a double slit on the upper surface of the leaf generally among the young growth, in which she inserts a double row of elongate eggs, which, as they swell, form a regular blister, but the most remarkable fact in the life history of this insect is the care she takes after laying her eggs. Nearly all insects after the eggs are laid leave them to their fate, but Lewis' saw-fly not only stands guard over them until they are hatched but further looks after the helpless grubs for some time after they have commenced feeding. She straddles the eggs with her wings half opened, the tip of her abdomen turned up, and with her jaws open, makes a slight buzzing sound if meddled with; if you pick her up, she never attempts to fly, but crawls back to her post, reminding one of an old hen protecting her chicks. The grubs when full grown are slightly under 1½ inches in length, general color dull brown to dirty yellow, covered with short brown hairs, the last abdominal segment yellow. When full grown they crawl into the ground and form the typical form of cocoon, generally in regular rows." Concerning the habits of the larvæ of another species (*Perga dorsalis*) he says (1918): " the gregarious larvæ feed at night, and rest during the day, clustered together in an oval mass, on the stem of the gum-tree upon which they are feeding. When disturbed, they exude a sticky yellow substance from the mouth, at the same time raising the tip of the body, and tapping it down on the foliage. The leaves are devoured from the top of the young gum trees; and when the larvæ are full fed, they crawl down the stem to pupate. I have found them fully developed in the middle of April; but when they descend from their resting place, they wander about over the grass for several days 'before they finally select a place in which to pupate, generally the softer soil against a tree-trunk. Into this they burrow to a depth of three or four inches, massing their large, oval cocoons in rows, one against the other. I watched several large swarms feeding upon the Peppermint-gums (*Eucalyptus novæ-angliæ*) at our Experiment Station at Uralla, and afterwards in their erratic wanderings over the grass; and marked down their final resting place and dug up the cocoons. At Binalong, in April, I observed two large swarms marching in massed formation; the heads of the hind rows always rested upon those in front as they moved along steadily together. Every now and then, the front rank came to a dead stop, when they all rested for three or four minutes; then a number began raising up and tapping down the tip of the abdomen, whereupon the whole band took up the motion; the leading ranks made a fresh start, and all moved along again. In the largest band, I counted two hundred and fifty caterpillars."

Recently Dr. Mann and I have observed bands of very similar saw-fly larvæ in British Guiana and Bolivia (cf. **Wheeler, W. M.**, and **Mann, W. M.**, A Singular Habit of Saw-fly Larvæ. Psyche 30, 1923, p. 9–13, 1 fig.).

2. **Bergson, H.**, L'Évolution Créatrice. Paris, Alcan 4th ed., 1908.

3. **Reuter, O. M.**, Lebensgewohnheiten, etc. (*loc. cit.*), p. 53.

4. For accounts of this behavior the following papers may be consulted: **Doten, S. B.**, Concerning the Relation of Food to Reproductive Activity and Longevity in Certain Hymenopterous Parasites. Tech. Bull. 78, Agr. State Univ. Nevada 1911, p. 7–30, 10 pls.; **Howard, L. O.**, On the Habit with Certain Chalcidoidea of Feeding at Puncture Holes made by the Ovipositor. Journ. Econ. Ent. 3, 1910, p. 257–260; **Lichtenstein, J. L.**, Sur la Biologie d'un Chalcidien. C. R. Acad. Sc. Paris 1921; **Lichtenstein, J. L.**, and **Rabaud, E.**, Le Comportement des "Polysphincta," Ichneumonides parasites des Araignées Bull. Biol. France Belg. 55, 1922, p. 267–287, 11 text-figs.; **Marchal, P.**, Observations biologiques sur un parasite de la Galéruque de l'Orme, *Tetrastichus xanthomelœnœ.* Bull. Soc. Ent. France 1905; La Ponte des Aphélines et l'intérêt individual dans les actes liés à la conservation de l'espèce. C. R. Acad. Sci. Paris, 148 p. 1223–1225, 1909; **Picard, F.**, Sur la biologie de *Tetrastichus rapo.* Bull. Soc. Ent. France 1921; **Picard, F.**, and **Rabaud, E.**, Sur le parasitisme externe des Braconides. Bull. Soc. Ent. France 1914, p. 266–269; **Roubaud, E.**, Observations biologiques sur *Nasonia brevicornis.* Bull. Biol. France Belg. (7) 50, 1917, p. 425–439, 1 fig. **Trouvelot, B.**, Observations biologiques sur l'*Habrobracon johansenii.* Soc. Biol. 1921. **Whiting, P. W.**, Sex-Determination and Biology of a Parasitic Wasp, *Habrobracon brevicornis* (Wesmael). Biol. Bull. 34, 1918, p. 250–256; Rearing Meal Moths and Parasitic Wasps for Experimental Purposes. Heredity 12, 1921 p. 255–261, 20 figs.

5. A family of great interest in connection with the evolution of both the solitary and social wasps is the Bethylidæ, since it shows transitions from parasitoidal to subsocial forms like Scleroderma, a

genus which, as will be shown below, should have been included in my list on page 18. The following literature deals with the habits of the Bethylidæ: **Bridwell, J. C.**, A Note on an Epyris and its Prey. Proc. Hawaiian Ent. Soc. (1916) 3, 1917, p. 262–263; Notes on the Bruchidæ and their Parasites in the Hawaiian Islands. *ibid.*, 1918, pp. 465–505; Some Notes on Hawaiian and Other Bethylidæ (Hymenoptera) with Descriptions of New Species. *ibid.* (1918) 4, 1919, p. 21–38; Some Notes on Hawaiian and Other Bethylidæ (Hymenoptera) with the Description of a New Genus and Species (second paper). *ibid.* (1919) 4, 1920, p. 291–341; **Ferton, Ch.**, *Perisemus 3-areolatus* Först. Notes detachées, etc. *l.c.*, 1901, p. 144; **Haliday, H. H.**, Notes on the Bethyli and on *Dryinus pedestris*. Ent. Mag. 2, 1835, p. 219–220; **Nielsen, I. C.**, Om *Perisemus fulvicornis* Gerst. En Overgangsform mellen Snylte-og Gravehvepsene. Ent. Meddel. (2) 2, 1903–'05, p. 105; **Williams, F. X.**, *Epyris extraneus* Bridwell (Bethylidæ), a Fossorial Wasp that preys on the Larva of the Tenebrionid Beetle *Gonocephalum seriatum* (Boiduval). Proc. Hawaiian Ent. Soc. 4, 1918, p. 55–63, 2 pls.

The studies of Bridwell on various species of Scleroderma show a distinct and very significant approach to social conditions. Both sexes of these small insects are dimorphic, each being represented by a winged and an apterous form, but the winged female and apterous male are much rarer than the apterous female and winged male. The normal prey is a beetle larva (Cerambycid, Bruchid) feeding in cavities in plants, but almost any firm, naked beetle, hymenopterous or dipterous larva of proper size may be substituted in the laboratory. After the female has found such a larva she attacks it though it may be several thousand times her own size and stings it repeatedly till it is completely paralyzed. Then she feeds for several days on its juices by biting it in various places and imbibing the exuding juices. After her eggs have matured she scatters them over the surface of the prey. This is devoured by the larvæ which soon after hatching assume an erect posture like that of the larval Epyris in Williams' figure (see page 51, Fig. 19 E). The female is long-lived and may remain with the larvæ and even mate with her own sons. Mr. Bridwell kindly gave me some cocoons of a Texan species, *Scleroderma macrogster* Ashm., which he had been rearing. During the summer of 1922 I bred four generations of this insect, using hickory and sumach borers (Clytinæ) and other larvæ (Pissodes, Tenebrionid and ant-larvæ) as prey. I was able not only to confirm all of Mr. Bridwell's interesting observations on the Hawaiian species of the genus but also to notice that the female may show an interest in her larvæ while they are feeding. She licks them assiduously at times and seems to feed on the juices which exude from the prey around their heads. This is interesting in connection with the discussion on trophallaxis in Belonogaster (see pp. 82–84). Moreover, several females readily co-operate in paralyzing the same beetle larva and feed and oviposit together upon it without the slightest signs of mutual antagonism. The recently emerged males eagerly gnaw their way into the silken cocoons and fecundate the females while they are still in the pupal stage. Scleroderma is therefore adelphogamic, and since a single male will mate with several females and a single female with several males, it is also polygamous and polyandric.

The peculiar Chalcidoid genus Melittobia is another group of para-sitoids, whose behavior, in certain particulars, recalls that of Sclero-derma. See the observations of S. I. Malyshev, Zur Biologie der Odynerus–Arten und ihrer Parasiten. Hor. Soc. Ent. Ross. 40, 1911, p. 1–58, 20 figs. Russian with Germ. résumé). The habits of this insect have also been studied by one of my students, Dr. L. H. Taylor, but his observations are still unpublished.

6. I give here a rather voluminous bibliography comprising several of the more important and more recent general works on the behavior of Sphecoid wasps, followed by a selection of papers dealing with the species of the various families and subfamilies. Many of the general works also contain materials on Vespoids.

SPHECOIDEA. (General). Adlerz, G., Biologiska meddelanden om Rofsteklar. Ent. Tidskr. 21, 1901, p. 161–200; Lefnadsförhållenden och Instinkter inom Familjerna Pompilidæ och Sphegidæ. K. Svensk. Vetensk. Akad. Handl. 1st part, 37, 1903, p. 1–181, 2nd part, 42, 1906, p. 1–48, 1 fig.; 3rd part, 45, 1910, p. 6–75; 4th part, 47, 1912, p. 5–61; Iakttgelser öfver solitära getingar, Arkiv. Zool. 3, 1907, 64 pp., 1 fig.; Orienteringsformägen hos steklar. Sundvall 1909; Ashmead, W. H., The Habits of the Aculeate Hymenoptera. Psyche 7, 1894, p. 19–26, 39–46, 59–66, 75–79; Banks, N., Sleeping Habits of Certain Hymenop-tera. Journ. N. Y. Ent. Soc. 10, 1902, p. 209–214; Borries, H., Bidrag til danske Gravehvepses Biologi. Vidensk. Meddel. naturh. Foren. 1897, p. 1–143; Bradley, J. C., A Case of Gregarious Sleeping Habits among Aculeate Hymenoptera. Ann. Ent. Soc. Amer. 1, 1908, p. 127–130; Brues, C. T., On the Sleeping Habits of Some Aculeate Hymen-optera. Journ. N. Y. Ent. Soc. 11, 1903, pp. 228–230; Descy, A., Observations sur le retour au nid des Hyménoptères. Bull. Soc. Ent. Belg. 4, 1922, p. 93–99, 104–111; Fabre, J. H., Études sur l'instinct et la métamorphose de Sphégiens. Ann. Sc. Nat. Zool. (4) 6, 1856, p.

137–189; Souvenirs Entomologiques. Paris, Delagrave, 10 series or volumes. (I have been able to ascertain only the following dates of publications: Vol. I, 1879, II, 1882, III, 1886, IX, 1905. The various chapters relating to the solitary wasps have been brought together and translated by A. T. de Mattos as " The Hunting Wasps " (1905) and " The Mason Wasps." (1919), New York, Dodd, Mead & Co.; Ferton, Ch., Notes Détachées sur l'instinct des Hyménoptères Mellifères et Ravisseurs. 1st series, Ann. Soc. Ent. France 70, 1901, p. 83–148, 3 pls.; 2nd ser. *ibid.* 71, 1902, p. 499–529, 1 pl.; 3rd ser. *ibid.* 74, 1905, p. 56–104, 2 pls.; 4th ser. *ibid.* 77, 1908, p. 535–584, 1 pl.; 5th ser. *ibid.* 78, 1909, p. 401–422; 6th ser. *ibid.* 79, 1910, p. 145–178; 7th ser. *ibid.* 80, 1911, p. 351–412; 8th ser. *ibid.* 83, 1914, p. 81–119, 3 pls.; 9th ser. *ibid.* 89, 1920–21, p. 329–375; Hyménoptères nouveaux d'Algérie et Observations sur l'Instinct d'une Espèce. Bull. Soc. Ent. France 1912, p. 186: De Gaulle, J., Catalogue Systématique et Biologique des Hyménopteres de France, p. 1–171, extr. de la Feuille de Jeune Natural. 1906, 1907 and 1908; Giraud, J., Mémoire sur les Insectes qui habitent les tiges séches de la Ronce. Ann. Soc. Ent. France (4) 6, 1866, p. 443–500; Hartman, C. G., Observations on the Habits of Some Solitary Wasps of Texas. Bull. Univ. Tex. Sc. Ser. 6, 1905, p. 1–72, 4 pls.; Nielsen, J. C., Biologiske Studier over Gravehvepse. Vidensk. Meddel. Naturh. Foren, 1900, p. 255; Biologische Studien über einige Grabwespen und solitäre Bienen. Allgem. Zeitschr. Ent. 6, 1901, p. 307, 308, 1 fig.; Peckham, G. W., and E. G., On the Instincts and Habits of the Solitary Wasps. Wis. Geol. Nat. Hist. Survey Bull. 2, Sc. Ser. 1, 1898, p. IV–245, 14 pls.; The Instincts of Wasps as a Problem in Evolution. Nature 59, 1899, p. 466–468, 2 figs.; Additional Observations on the Instincts and Habits of the Solitary Wasps. Bull. Wis. Nat. Hist. Soc. 1, 1900, p. 85–93; Wasps, Social and Solitary. Westminster, 1905; Rabaud, E., L'Instinct paralyseur des Hymenopteres vulvérants. C. R. Acad. Ac. Paris 165, 1917, p. 680; Le Contraste entre le régime alimentaire des larves et celui des adultes chez divers insectes. Bull. Biol. France Belg. 56, 1922, p. 230–243; Rau, P. and N., The Sleep of Insects, an Ecological Study. Ann. Ent. Soc. Amer. 9, 1916, p. 227–274; Wasp Studies Afield. Princeton Univ. Press. 1918; Ecological and Behavior Notes on Missouri Insects. Trans. Acad. Sc. St. Louis 24, 1922, 71 pp., 4 pls.; Read, C., Instinct, especially in Solitary Wasps. Brit. Journ. Psychol. 4, 1911, p. 1–32; Reuter, O. M., Lebensgewohnheiten u. Instinkte, etc. 1913. *loc. cit.*; Roubaud, E., Le venin et l'évolution paralysante chez les Hyménoptères prédateurs. Bull. Biol. France Belg. 51, 1917, p. 400–419; Rudow, F., Die Wohnungen der Raub-, Grab- und Faltenwespen.

Sphegiden, Crabroniden, Vespiden. Perleberg, 1905; **Verhoeff, C.,** Beiträge zur Biologie der Hymenoptera. Zool. Jahrb. Syst. 6, 1892, p. 680–754, 2 pls., 7 text-figs.; Neue und wenig bekannte Gesetze aus der Hymenopteren-Biologie. Zool. Anzeig. 15, 1892, p. 362–370; **Williams, F. X.,** Philippine Wasp Studies. Bull. No. 14, Exper. Station Hawaiian Sugar Plant. Assoc. 1919, 186 pp., 106 text-figs.

AMPULICIDAE. **Ferton, C.,** Sur les Moeurs du *Dolichurus haemorrhous* Costa. Act. Soc. Linn. Bordeaux (5) 7, 1894, p. 215–221, 1 pl.; **Picard, F.,** Sur les Mœurs et le genre de proie de l'*Ampulex fasciatus* Jurine. Bull. Soc. Ent. France 1911, p. 113–116, 2 figs. **de Réaumur, R. A. F.,** Mémoires 6, 1742, pp. 280 *et seq.,* Pl. 26–28 (Cossigni's observations on Ampulex in Mauritius); **Williams, F. X.,** *Dolichurus stantoni.* Phil. Wasp Studies, *l.c.,* p. 111–117.

SPHECIDAE, subfam. NYSSONINAE. **Barth, G. P.,** Observations on the Nesting Habits of *Gorytes canaliculatus* Pack. Bull. Wis. Nat. Hist. Soc. N. S. 5, 1907, p. 141–149, 4 figs.; **Dahlbom. A. G.,** Oplysninger angaande *Diodonti tristis* og *Alysonii Ratzeburgii* Levemaade. Förh. Skand. Naturf. 4, Möde, Christiania (1844) 1847, p. 277–280; **Ferton, C.,** Gorytes. Notes detach. *loc. cit.* 4, p. 558; *ibid.* 6, p. 151–152, 158–159; *ibid.* 7, p. 369; Didineis. *ibid.* 7, p. 404–406; **Rau, P.,** and **N.,** *Alyson melleus.* Wasp Studies afield. *loc. cit.* Chapt. 9.

PSENINAE. **Barth, G. P.,** On the Nesting Habits of *Psen barthi* Vier. Bull. Wis. Nat. Hist. Soc. 5, 1907, p. 251–257, 9 figs.; **Bordage, E.,** Notes biologiques recueillies a l'île de la Reunion. Bull. Sc. France Belg. (7) 46, 1912, p. 29–83, 2 pls., 7 figs. (Pison); **Peckham, G. W.** and **E. G.,** Solitary Wasps. *l.c.,* Chapt. 3 (Stigmus); **Williams, F. X.,** Some Observations on the Leaf-Hopper Wasp, *Nesomimesa hawaiiensis* Perkins, at Papala, Hawaii. Proc. Hawaiian Ent. Soc. 4, 1918, p. 63–68, 3 figs.

OXYBELINAE. **Ferton,** Notes détach *l.c.* 1, 1901, p. 110–112, (Oxybelus); *ibid.* 2, p. 516–518; *ibid.* 4, p. 564; **Gerstaecker, C. E. A.** Ueber die Gattung Oxybelus Latr. Zeitschr. ges. Naturw. Berlin, 1867; **Kieffer, J. J.,** Zur Lebensweise v. *Oxybelus uniglumis* Dahlb. u. ihrer Parasiten. Allg. Zeitschr. Ent. 7, 1902, p. 81–84; **Peckham, G. W.** and **E. G.,** Solitar. Wasps *l.c.* Chapt. 7.

CRABRONINAE. **Barth, G. P.,** The Nesting of *Anacrabro ocellatus* Pack. Bull. Wis. Nat. Hist. Soc. 6, p. 147–153, 3 pls., 4 figs.; **Emery, C.,** Sur un Crabronide chasseur de fourmis. Bull. Soc. Ent. France, 1893, p. LXIII–LXIV; **Enslin, E.,** Zur Biologie des *Solenius rubicola* Douf. et Perr. (*larvatus* Wesm.) und seiner Parasiten. Konowia 1, 1922, p. 1–16, 7 figs.; **Ferton, C.,** Un Hyménoptère ravisseur de fourmis. Act. Soc. Linn. Bordeaux (5) 4, 1890, p. 341–346;

Notes détach. *l.c.* 2, p. 518–519, 3, p. 71–73 (Crabro); **Mally, C. W.,** Bee Pirates. Agric. Journ. Cape Good Hope 33, 1908, p. 206–213, 4 figs.; **Marchal, P.,** Observations sur les Crabronides. Ann. Soc. Ent. France 1893, p. 331–338, 1 pl.; **Nielsen, J. C.,** Jagttagelser över nogle danske Garvehvepses Biologi. Ent. Meddel. (2) 2, 1903, p. 110–114; **Peckhams,** Solitary Wasps. *l.c.* Chapt. 3 and 4; **Rau, P.** and **N.,** Wasp Studies. *l.c.* Chapt. 3; **Rudow, F.,** Die Lebensgewohnheiten der Crabronen. Insektenborse 14, 1897, p. 255, 261; **Waga, A.,** Instinct des Hyménoptères Crabroniens. Le Naturaliste, 15 Mars, 1882.

PHILANTHINAE. **Ainslie, C. N.,** A Note on the Habits of Aphilanthops. Canad. Ent. 41, 1909, p. 99–100; **Bouvier, E. L.,** Les variations des habitudes chez les Philanthes. C. R. Soc. Biol., Paris 52, 1901, p. 1129–1131; **Fabre, J. H.,** Souv. Ent. *l.c.* 4, Chapt. 11; **Ferton, C.,** Notes détach. *l.c.* 3, p. 66–67; **Peckham, G. W.** and **E. G.,** Solitary Wasps. *l.c.*, Chapt. 11; **Picard, F.,** Note sur l'instinct de Philanthe apivore Feuille jeune Natural. (4) 34, 1903, p. 17; **Rau, P.** and **N.,** Wasp Studies. *l.c.*, Chapt. 5; **Wheeler, W. M.,** A Solitary Wasp (*Aphilanthops frigidus* F. Smith) that Provisions its Nest with Queen Ants. Journ. Anim. Behav. 3, 1913, p. 374–387.

TRYPOXYLONINAE. **Bordage, E.,** Notes Biologiques etc. *l.c.;* **du Buysson, R.,** Le nid et la larve du *Trypoxylon albitarse* F. Ann. Soc. Ent. France 67, 1898, p. 84–86, 2 pls.; **Ferton,** Notes détach. *l.c.* 4, p. 563; 6, p. 155–156; **Green, E. E.,** On the Nesting Habits of *Trypoxylon intrudens* and *Stigmus niger.* Spol. Zeylanica 1, 1903, p. 68–70, 2 figs.; **Kleine, R.,** Zwei merkwürdige Nestanlagen v. *Trypoxylon figulus* L. Zeitschr. wiss. Insektenbiol. 6, 1910, p. 24–25; **Peckham, G. W.** and **E. G.,** Notes on the Habits of *Trypoxylon rubrocinctum* and *T. albopilosum.* Psyche, 1895, p. 303–306; also Solitary Wasps. *l.c.*, Chapt. 7; **Popovici-Baznosanu, A.,** Contribution à l'étude des Sphegiens (Trypoxylon et Psenulus). Arch. Zool. Expér. (5) 6, 1911; **Rau, P.** and **N.,** Wasp Studies. *l.c.*, Chapt. 8; **Turner, C. H.,** Reactions of the Mason Wasp *Trypoxylon albitarsus* to light. Journ. Animal. Behav. 2, 1912, p. 353–362, 5 figs.; **Williams, F. X.,** Philip. Wasp Stud. *l.c.*, p. 142–145.

MELLININAE. **Lucas, H.,** Quelques Remarques sur la Maniere de vivre du *Mellinus sabulosus.* Ann. Soc. Ent. France (4) 1, 1861, p. 219; **Rabaud, E.,** Note sur l'instinct de *Mellinus arvensis* et ses rapports avec celui des autres Sphégiens. Bull. Biol. France. Belg. 51, 1917, p. 331–346; **Steinvorth, H.,** Eine Raubwespe (*Mellinus arvensis*). Jahresb. naturw. Ver. Lüneburg 3, 1867, p. 142–144; **Verhoeff, C.,** Beiträge zur Biologie verschiedener Hymenopteren-Arten, *Mellinus arvensis.* Zool. Jahrb. Biol. 6, 1892, p. 696–699.

SPHECINAE. **Adlerz, G.,** Nya iakttagelser öfver *Ammophila (Miscus)* *campestris.* Ent. Tidskr. 30, 1909, p. 163–176; **Descy, A.,** Instinct et Intelligence. Expériences sur l'Ammophile. Ann. Soc. Ent. Belg. 59, 1919, p. 86–95; L'Ammophile du Sable. Bull. Soc. Ent. Belg. 7, 1919, p. 123–132; *ibid.* 8, 1919, p. 136–142; *ibid.* 9, 1919, p. 147–158. **Fabre, J. H.,** Souvenirs Ent. *l.c.* 1, Chapt. 6–12 (Sphex); 4, Chapt. 1–5 (Pelopœus); 1, Chapt. 14, 15 and 19, 2, Chapt. 2–4 and 4, Chapt. 12 (Ammophila); **Ferton, C.,** Notes détach. *l.c. passim* (Sphex, Sceliphron, Ammophila); **Hubbard, H. G.,** Some Insects which brave the Dangers of the Pitcher Plant. Proc. Ent. Soc. Wash. 3, 1895, p. 314, 315; **Jones, F. M.,** Pitcher Plant Insects. Ent. News 1904, p. 14–17, 2 pls.; **Kirchner, A.,** Zur Naturgeschichte der *Ammophila arenaria* Dahlb. Lotos 8, 1858, p. 85–87; **Lüderwaldt, H.,** *Sphex striatus* Sm., bei seinem Brutgeschäft. Zeitschr. wiss. Insektenbiol. 6, 1910, p. 177–179; **Maindron, M.,** Notes pour servir a l'histoire des Hyménoptères de l'Archipel Indien et de la Nouvelle-Guinée. Ann. Soc. Ent. France (5) 8, 1878, p. 385–398, 1 pl.; *ibid.* (5) 9, 1879, p. 173–182, 1 pl.; **Marchal, P.,** Études sur l'instinct de *l'Ammophila affinis.* Arch. Zool. Expér. Gen. (2) 10, 1892, p. 23–36; **Peckham, G. W.,** and **E. G.,** Solitary Wasps. *l.c.* Chapt. 1, 2, 13 and 14 (Ammophila, Sphex, Priononyx, Chlorion, etc.); **Pergande, T.,** Peculiar Habit of *Ammophila gryphus* Sm. Proc. Ent. Soc. Wash. p. 256–258, 1 fig.; **Picard, F.,** Mœurs de *l'Ammophila tydei.* Feuille jeun. Natural. (4) 34, 1903, p. 15–17; Recherches sur l'éthologie du "*Sphex Maxillosus*" F. Mém. Soc. Nat. Sc. nat. math. Cherbourg 33, 1903, p. 97–130; **Rabaud, E.,** Observations et expériences sur *Ammophila heydeni.* Bull. Soc. Zool. France 1919; **Rau, P. and N.,** The Biology of the Mud-daubing Wasps as Revealed by the Contents of their Nests. Journ. Anim. Behav. 6, 1916, p. 27–63, 5 pls.; Wasp Studies. *l.c.* Chapt. 6, 10 and 11; **Schoenichen, W.,** Die grosse gelbe Grabwespe, *Sphex ichneumonea.* Prometheus 13, 1902, p. 777–780, 1 fig.; **Scholz, E. J. R.,** Die Lebensgewohnheiten schlesischer Grabwespen II, Zeitschr. Insektenbiol. 5, 1909, p. 179–182, 2 figs.; **De Stefani, T.,** Sulla nidificazione e biologia dello *Sphex paludosus* Rossi. Natural. Sicil. N. S. 1, 1896, p. 131–136; Ulteriori osservazioni sulla nidificazione dello *Sphex paludosus.* Monit. zool. Ital. 12, 1901, p. 222–223; **Schirmer, C.,** *Psammophila viatica* L. Ill. Zeitschr. Ent. 3, 1898, p. 265; **Turner, C. H.,** The Homing of the Mud-Dauber. Biol. Bull. 15, 1908, p. 215–225; The Copulation of *Ammophila abbreviata.* Psyche 19, 1912, p. 137, 1 fig.; A Note on the Hunting Habits of an American Ammophila. *ibid.* 18, 1911, p. 13, 14; Sphex Overcoming Obstacles. *ibid.* 19, 1912, p. 100–102; **Walsh, B.,** The Digger Wasps, Amer. Ent. 1, 1869, p. 123; **Williams, F. X.,** Philip. Wasp Studies *l.c.*, p. 117–131 (Sphecinæ); **Williston, S. W.**, Notes on the Habits of Ammophila. Ent. News 3, 1892, p. 85, 86.

LARRINAE. **Fabre, J. H.**, Souvenirs Ent. *l.c.* 3, Chapt. 12 (Tachytes); **Ferton, C.**, Notes détach. *l.c.*, *passim;* **Rau, P.** and **N.**, Wasp Studies *l.c.*, Chapt. 8 and 9; **Vincens, F.**, Observations sur les moeurs et l'instinct d'un insecte hyménoptère, le *Nitela spinolai.* Bull. Soc. Hist. Nat. Toulouse 43, 1910, p. 11–18; **Williams, F. X.**, Monograph of the Larridæ of Kansas. Kans. Univ. Sc. Bull. 8, 1913, p. 121–213, 9 pls.; Philip. Wasp Studies *l.c.*, p. 131–142 (Larrinæ); **Xambeu, V.**, Description des premiers états du *Sylaon xambeui* E. André, du groupe des Larrides. Bull. Soc. Ent. France 1896, p. 79–80.

ASTATINAE. **Peckham, G. W.** and **E. G.**, Solitary Wasps. *l.c.*, Chapt. 9 (*Astata unicolor* and *bicolor*).

PEMPHREDONINAE. **Davidson, A.**, Habits of *Stigmus inordinatus.* Psyche 7, 1895, p. 271; **Dahlbom, A. G.**, Oplysminger, etc., *l.c.* (Diodontus); **Ferton, C.**, Notes détach. *l.c.*, 3, p. 73 and 4, p. 563 (Diodontus; **Green, E. E.**, On the Nesting Habits of *Trypoxylon intrudens* and *Stigmus niger. l.c.;* **Peckham, G. W.**, and **E. G.**, Solitary Wasps. *l.c.*, Chapt. 10 (*Diodontus americanus*); **Rau, P.** and **N.**, Wasp Studies. *l.c.*, Chapt. 4 (Diodontus and Pemphredon = Xylocelia and Ceratophorus).

CERCERIDAE. **Alfken, J. D.**, Ueber das Leben von *Cerceris arenaria* L. und *rybiensis* L. Ent. Nachr. 25, 1899, p. 106–111. **Dufour, L.**, Observations sur les métamorphoses du *Cerceris bupresticida* et sur l'industrie et l'instinct entomologique de cet Hyménoptère. Ann. Sc. Nat. (2) 15, 1841, p. 353–370, fig.; **Fabre, J. H.**, Observations sur les moeurs des Cerceris et sur la longue conservation des Coléoptères dont ils approvisionnent leurs larves. *ibid.* (4) 4, 1855, p. 129–150; Souvenirs Ent. *l.c.* 1, Chapt. 3–5; **Ferton, C.**, Notes détach. *l.c.* 1, p. 109; 3, p. 65–66; 6, p. 153–155; **Grossbeck, J. A.**, Habits of *Cerceris fumipennis* Say. Journ. N. Y. Ent. Soc. 20, 1912, p. 135; **Marchal, P.**, Études sur l'Instinct de *Cerceris ornata.* Arch. Zool. Expér. Gén. (2) 5, 1887, p. 27–60, 6 figs.; **Nielsen, J. C.**, Zur Lebensweise und Entwickelung von *Ceratocolus subterraneus* Fabr. Zeitschr. Ent. 7, 1902, p. 178–180, 2 figs.; **Peckham, G. W.** and **E. G.**, Solitary Wasps. *l.c.* Chapt. 11; **Rau, P.** and **N.**, Wasp Studies. *l.c.*, Chapt. 8; **Williams, F. X.**, Philip. Wasp Studies. *l.c.*, 145–149 (Cerceridæ).

BEMBICIDAE. **Bouvier, E. L.**, Le retour au nid chez les Hyménoptères prédateurs du genre Bembex. C. R. Soc. Biol. Paris, 52, 1900, p. 874–876; Les habitudes des Bembex. Ann. Psych. 7, 1901, p. 1–68, 4 figs.; **Brèthes, F. J.**, Notes biologiques sur trois Hyménoptères de Buenos Ayres. Rev. Mus. la Plata 10, 1902, p. 195–205, 1 pl. (Monedula); **Dufour, L.**, Observations sur le genre Stizus. Ann. Soc. Ent.

300   SOCIAL LIFE AMONG THE INSECTS

France 7, 1838, p. 269–279, fig.; **Fabre, J. H.,** Souvenirs Ent., *l.c.* 1, Chapt. 16–18; **Ferton, C.,** Observations sur l'instinct des Bembex Fabr. Act. Soc. Linn. Bordeaux 54, 1899, p. 331–345; Sur les mœurs du *Stizus fasciatus* Fabr. C. R. Ass. franç. Av. Sc. 30me Sess. 1re part, p. 152, 2me part, p. 680–683; Notes détach. *l.c. passim* (Bembex, Stizus); **Handlirsch, A.,** Monographie der mit Nysson und Bembex verwandten Grabwespen. SB. K. Akad. Wiss. Wien, 8 parts, 1887–1895; **Hine, J. S.,** A Preliminary Report on the Horse-flies of Louisiana, with a Discussion of Remedies and Natural Enemies. Circ. No. 6, State Crop Pest Com. La. 1906, p. 7–43, 20 figs. (*Stictia carolina*); **Jacobson, E.,** Observations sur les habitudes du *Bembex Borrei* Handl., Lettre addressée de Batavia à M. le Prof. Bouvier. Bull. Mus. d'Hist. Nat. Paris, 1909, p. 451–453; **Marchal, P.,** Remarques sur les Bembex. Ann. Soc. Ent. France 62, 1893, p. 93–98; **Marchand, E.,** Sur le retour au nid du *Bembex rostrata* Fabr. (unique observation). Bull. Soc. Sc. Nat. Ouest Nantes 10, 1901, p. 247–250; **Melander, A. L.,** How Does a Wasp Live at Home? State College Bull. Pullman, Wash. 3, 1904, 4 pp., 11 figs. (Bembex); **Parker, J. B.,** Notes on the Nesting Habits of *Bembex nubilipennis.* Ohio Natural., May 1910, p. 163–165; Notes on the Nesting Habits of Some Solitary Wasps. Proc. Ent. Soc. Wash. 17, 1915, p. 70–77, 1 pl.; A Revision of the Bembecine Wasps of America North of Mexico. Proc. U. S. Nat. Mus. 52, 1917, 155 pp., 8 pls.; **Peckham, G. W.** and **E. G.,** Solitary Wasps *l.c.,* Chapt. 6; **Rau, P.** and **N.,** Wasp Studies *l.c.,* Chapt. 1 (Bembex); Chapt. 10 (Relation of *Stizus unicinctus* to *Priononyx thomæ*); **Riley, C. V.,** The Larger Digger Wasp. Insect Life 4, 1892, p. 248–252, 7 figs. (*Sphecius speciosus*); **Roubaud, E.,** Bembex chasseur de Glossines au Dahomey. C. R., Acad. Paris 151, 1910, p. 505–508; **Sergent, E.** and **Et.,** A propos d'un essai d'acclimatement des Monedula en Algérie. Bull. Soc. Hist. Nat. Afr. Nord. 2, 1910, p. 81–82; **Schoenichen, W.,** Die Lebensgewohnheiten der Wirbelwespe (*Bembex spinolæ*). Prometheus 15, 1904, p. 761–764, 2 figs.; **Schuster, W.,** Aufzeichnungen über *Bembex rostrata,* die grösste deutsche Mordwespe. Wien. Ent. Zeitg. 27, 1908, p. 124–126; **Siebertz, C.,** *Bembex rostrata* L. Nerthus 5, 1903, p. 421–423, 449–451, 1 pl., 8 figs.; **Wesenburg-Lund, C.,** *Bembex rostrata,* dens Liv og Instinkter. Ent. Meddel. 3, 1891, p. 1–26.

7.   The following list of literature on the Solitary Vespoids comprises only a small portion of what has been published on these insects:

Scoliidae.   **Adlerz, G.,** *Tiphia femorata* Fabr. dess levnadssätt och utvecklingsstadier. Ark. Zool. 7, 1911; **Davis, J. J.,** Contributions to a Knowledge of the Natural Enemies of Phyllophaga. Bull. Nat. Hist. Survey Illinois 13, 1919, p. 53–138, 13 pls., 46 text-figs.; **Davis, J. J.,** and **Luginbill, P.,** The Green June Beetle or Fig Eater. Bull. 242, N. Carolina Agri. Exper. Station 1921, 35 pp., 9 figs. (*Scolia dubia* Say); **Fabre, J. H.,** Souvenirs Ent. *l.c.* 3, Chapt. 1–4; 4, Chapt. 13; **Ferton, Ch.,** Notes Détach. *l.c.* 7, p. 409–411 (*Myzine andrei*); **Flint, W. P.,** and **Sanders, G. E.,** Note on a Parasite of White Grubs. Journ. Econ. Ent. 5, 1912, p. 490 (*Myzine 6-cincta*); **Forbes, S. A.,** On the Life-History, Habits and Economic Relations of the White Grubs and May-Beetles (Lachnosterna). 24th Rep. State Ent. Nox. Benef. Ins. Illinois 1908, p. 135–168, 3 pls. (*Tiphia vulgaris*); **Froggatt, W. W.,** A Natural Enemy of the Sugar Cane Beetle of Queensland. Agric. Gaz. N. S. Wales 13, 1902, p. 63–68, 1 pl. (*Dielis formosa*); **Muir, F.,** The Introduction of *Scolia manilæ* Ash. into the Hawaiian Islands. Ann. Ent. Soc. Amer. 10, 1917, p. 207–210; **Nowell, W.,** Two Scoliid Parasites on Scarabæid Larvæ in Barbados. Ann. Applied Biol. 2, 1915, also West. Ind. Bull. 15, p. 149–158, 1 pl.; **Passerini, C.,** Osservazioni sulle Larve, Ninfe ed Abitudini della *Scolia flavifrons*. Pisa 1840, 15 pp. 1 pl.; Continuazione delle Osservazioni sulle Larve della *Scolia flavifrons*. Firenze, Pozzati 1841, 7 pp., 1 pl.; **Rau, P. and N.,** Wasp Studies, *l.c.*, Chapt. 7.; **Smythe, E. G.,** The White Grubs injuring Sugar Cane in Porto Rico I. Life Cycles of the May Beetles or Melolonthids. Journ. Dep. Agric. Porto Rico, 1 and 2, 1917, p. 47–92, 9 pls. (Elis, Campsomeris and Tiphia); **Swezey, O. H.,** *Scolia manilæ*, a successfully introduced parasite for the Anomala grub. Hawaiian Plant. Rec. 17, 1917, p. 50–55, 5 figs.; **Williams, F. X.,** Philip. Wasp Studies, *l.c.*, p. 54–61 (*Scolia manilæ*), p. 61–69 (3 species of Tiphia); **Wolcott, G. N.,** Notes on the Life History and Ecology of *Tiphia inornata* Say. Journ. Econ. Ent. 7, 1914, p. 382–389.

THYNNIDAE. **Bridwell, J. C.,** Notes on the Thynnidæ. Proc. Hawaiian Ent. Soc. 3, 1917, p. 263–265; **Williams, F. X.,** A Note on the Habits of *Epactiothynnus opaciventris* Turner, an Australian Thynnid Wasp. Psyche 26, 1919, p. 160–162, 2 figs.

METHOCIDAE. **Adlerz, G.,** La proie de *Methoca ichneumonides* Latr. Ark. Zool. 1, 1904, p. 255–258; *Methoca ichneumonides* Latr., dess lefnadssätt och utvecklingsstadier. *ibid.* 3, 1906, 48 pp., 1 pl.; **Bouwman, B. E.,** Ueber die Lebensweise von *Methoca ichneumonides* Latr. Tijdschr. Ent. 52, 1909, p. 284–294, 296–299, 8 figs.; **Champion, H. G.** and **R. J.,** Observations on the Life-History of *Methoca ichneumonides* Latr. Ent. Month. Mag. 20, 1914, p. 266–270; **Champion, H. G.,** Addendum to Observations on the Life-History of *Methoca ichneumonides* Latr. *ibid.* 21, 1916, p. 40–42; **Williams, F. X.,** Notes on the Life-History of *Methoca stygia* Say. Psyche 23, 1916, p. 121–125, 1 pl.;

Philip. Wasp Studies *l.c.*, p. 69–79 (*Methoca striatella* and *punctata*).
MUTILLIDAE. **Borries, H.,** *Mutilla erythrocephala* Fabr., som parasit
hos *Crabro* (*Solenius*) *rubicola* D. and P. Ent. Tidskr. 13, 1892, p.
247; **Ferton, Ch.,** Notes détach. *l.c.* 4, p. 574 and 9, p. 359–363 (*Steno-
mutilla argentata*); 4, p. 573–574 (*Cystomutilla ruficeps*); **Rucker,
Miss A.,** A Glimpse of the Life History of *Mutilla vesta* Cresson. Ent.
News 14, 1903, p. 75–77.

PSAMMOCHARIDAE (Pompilidæ). **Adlerz, G.,** *Ceropales maculata* Fabr.
en parasitisk Pompilid. Bih. svenska Vet. Akad. Handl. 28, 1902,
20 pp.; Lefnadsförhållanden och Instinkter, etc., 1903–1912 *l.c.;* **von
Buttel–Reepen, H.,** Psychobiologische u. biologische Beobachtungen, etc.
Naturw. Wochenschr. n.f. 6, 1907 (Jacobson's observations on *Macro-
meris splendida*); **Davidson, A.,** An Enemy of the Trap Door Spider.
Ent. News 16, 1905, p. 233–234 (*Parapompilus planatus*); **Fabre, J. H.,**
Souvenirs Ent. *l.c.* 2, Chapt. 12; 4, Chapt. 2 and 14; **Ferton, Ch.,** Sur
les Mœurs de Miscophus. Act. Soc. Linn. Bordeaux 48, 1895, p. 266–
268; Nouvelles observations sur l'instinct des Pompilides. *ibid.* 52,
1897, p. 101–132; Notes détach. 1901–1921. *l.c. passim* (many species);
Johnson, S. A., Nests of *Agenia architecta* Say. Ent. News 14, 1903,
p. 290; **Laboulbène, A.,** Sur un Hyménoptère fouisseur du Genre Pepsis
qui approvisionne ses larves avec une grosse espèce de Mygale et
remarques sur quelques parasites des Araignées. Ann. Soc. Ent. France
1895, p. 179–190; **Marchal, P.,** Le Retour au Nid chez le *Pompilus
sericeus* V.d.L. C. R. Séances Soc. Biol., Dec. 22, 1900; **Needham, J. G.**
and **Lloyd, J. F.,** Life of Inland Waters, p. 333 (*Priocnemis flavicornis*);
Peckham, G. W., and **E. G.,** Solitary Wasps. *l.c.*, Chapt. 5 and 12;
Pérez, J., Notes Zoologiques. Act. Soc. Linn. Bordeaux (5) 47, 1894,
p. 231–331, 1 fig. (Ceropales); **Picard, F.,** Note sur l'instinct du
Pompilus viaticus. Feuille jeun. Natural. (4) 34, 1904, p. 142–145;
Rabaud, E., Notes critiques sur les mœurs des Pompilides. Bull. Soc.
France Belg. 43, 1909, p. 171–182; L'instinct Paralyseur des Araignées.
C. R. Acad. Sc. Paris 172, 1911, p. 289–291; **Rau, P.** and **N.,** Wasp
Studies. *l.c.*, Chapt. 2; **Schoenichen, W.,** Die Spinnenmörder (Pom-
piliden). Prometheus 15, 1904, p. 89–92, 4 figs.; **Wickham, H. F.,** Habits
of a Wasp (*Agenia architecta* Say). Ent. News 9, 1898, p. 47; **Williams,
F. X.,** *Philip.* Wasp Studies *l.c.*, 79–110 (Psammocharidæ).

MASARIDAE. **Brauns, H.,** Biologisches über südafrikanische Hymenopteren. Zeitschr. wiss. Insektenbiol. 6, 1910, p. 384–387, 445–447; 7,
1911, p. 16–19, 90–92, 117–120, 238–240; **Ferton, C.,** Notes détach. *l.c.*
1901, p. 137–139 (Ceramius and Celonites); *ibid.* 6, p. 174–176 (Celonites); *ibid.* 9, p. 372–374 (Masaris); **Giraud, J.,** Notes sur les Mœurs
du *Ceramius lusitanicus* Klug. Ann. Soc. Ent. France (5) 1, 1871,

p. 375–379; de Saussure, H. F., Note sur la tribu des Masariens et principalement sur le *Masaris vespiformis*. Ann. Soc. Ent. France (3) 1, 1853, p. 17–21.

EUMENIDAE. Adlerz, G., Iakttagelser öfver *Hoplomerus reniformis* Wesm. Ent. Tidskr. 23, 1902, p. 241–252; Borries, H., Om Redebygningen hos *Ancistrocerus oviventris* Wesm. Vid. Meddel. naturh. Foren 1897, p. 160–163, 1 fig.; Bequaert, J., A Revision of the Vespidæ of the Belgian Congo, based on the Collection of the American Museum Congo Expedition, with a List of Ethiopian Diplopterous Wasps. Bull. Amer. Mus. Nat. Hist. 39, 1918, p. 1–384, 267 figs., 6 pls.; Bruch, C., Le Nid de l'*Eumenes caniculata* (Oliv.) Sauss. (guêpe solitaire) et observations sur deux de ses parasites. Rev. Mus. La Plata 11, 1904, p. 223–226, 1 pl.; Cretin, E., Some Observations on *Eumenes dimidiatipennis*. Journ. Bombay Nat. Hist. Soc. 14, 1903, p. 820–824; Davidson, A., Habits and Parasites of a New Californian Wasp. Psyche 7, 1896, p. 335–336; Notes on California Wasps. The Nesting Habits of *Ancistrocerus birenimaculatus* Sauss. Ent. News 10, 1899, p. 180–181; Dufour, L., Mémoire pour servir a l'histoire de l'industrie et des métamorphoses des Odynères et description de quelques nouvelles espèces de ce genre des insects. Ann. Sc. Nat. (2) 11, 1839, p. 85–103, fig.; Fabre, J. H., Souvenirs Ent. *l.c.* 2, Chapt. 5 and 6; 4, Chapt. 10; Ferton, C., Observations sur l'Instinct des quelques Hyménoptères du Genre Odynerus. Act. Soc. Linn. Bordeaux 48, 1896, p. 219–220; Notes détach. *l.c. passim* (many species); Forbes, H. O., A Naturalist's Wanderings in the Eastern Archipelago. London 1885, p. 72–73 (*Zethus cyanopterus*); Forbes, S. A., An entomological train-wrecker (*Odynerus foraminatus* Saussure). 20th Rep. Nox. Benef. Insects Illinois 1898, p. 103–105, 1 pl.; Handlirsch, A., Fossile Wespennester. Ber. senkenb. Naturf. Gesell. Frankf. a. M. 1910, p. 265–266, 1 fig.; Hartman, C. G., The Habits of *Eumenes belfragei* Cress. Journ, Anim. Behav. 3, 1913, p. 353–360, 7 figs.; Howes, P. G., Tropical Life in British Guiana (Beebe). Ent. Part III, 1917, p. 371–450, 2 pls., 15 figs.; Hungerford, H. B., and Williams, F. X., Biological Notes on Some Kansas Hymenoptera. Ent. News 23, 1912, p. 241–260, 3 pls. Iseley, D., The Biology of Some Kansas Eumenidæ. Kans. Univ. Sc. Bull. 8, 1913, p. 235–309, 4 pls.; Kriechbaumer, Eumenidenstudien. Ent. Nachr. 5, 1879, p. 1–4, 57, 59, 85, 201, 204, 309–312; Laloy, L., Les Odynères, Naturaliste 27, 1905, p. 273–275; Les Eumènes. *ibid.* 28, 1906, p. 153–154; Maindron, M., Histoire des Guêpes Solitaires de l'Archipel Indien et de la Nouvelle Guinée. Ann. Soc. Ent. France (6) 2, 1882, p. 69–71, 169–188, 267–286, 2 pls.; Malyshev, S. I., Zur Biologie der Odynerus-Arten, etc. *l.c.*; Marlatt, C. L., Food–Habits of Odynerus. Proc. Ent. Soc. Wash. 6,

1894, p. 172–173. **Mjöberg, E.**, Biologiska iakttagelser öfver *Odynerus oviventris* Wesm. Ark. Zool. 5, 1909, 8 pp., 2 figs.; **Morice, F. D.**, Nidification of *Odynerus reniformis* Gmél. near Chobham. Ent. Month. Mag. (2) 17, 1906, p. 216–220; **Nietner, J.**, Beobachtungen über den Haushalt von *Eumenes Saundersii* Westw. Stett. Ent. Zeitg. 16, 1855, p. 223–226; **Perris, E.**, Notice sur les habitudes et les métamorphoses de l'*Eumenes infundibuliformis Oliv.*, *E. olivieri* St. Farg. Ann. Soc. Ent. France (2) 7, 1849, p. 185–194; Note additionelle sur les habitudes et les métamorphoses de l'*Eumenes infundibuliformis* Oliv. *ibid.* (2) 10, 1852, p. 557–559; **Poulton, E. B.**, Mr. W. A. Lamborn's Observation on Marriage by Capture by a West African Wasp. A Possible Explanation of the Great Variability of Certain Secondary Sexual Characters in Males. Rep. 83rd Meet. Brit. Assoc. Adv. Sc. 1914, p. 511–512; **Rau, P.** and **N.**, A Sleepy Eumenid. Ent. News 24, 1913, p. 396; Wasp Studies *l.c.*, Chapt. 13; **Roubaud, E.**, Gradation et perfectionnement de l'instinct chez les Guêpes solitaires d'Afrique, du genre Synagris. C. R. Acad. Sc. 147, 1908, p. 695–697; Recherches sur la Biologie des Synagris. Evolution de l'instinct chez les Guêpes Solitaires. Ann. Soc. Ent. France 79, 1910, p. 1–21, Transl. Ann. Rep. Smithson. Inst. 1910, p. 507–525, 4 pls.; **Saunders, S. S.**, On the Habits and Economy of Certain Hymenopterous Insects which nidificate in Briars and their Parasites. Trans. Ent. Soc. London 1873, p. 407 (*Raphiglossa zethoides*); Notes on the British Species of the Genus Odynerus. Ent. Month. Mag. 15, 1879, p. 249–250; **de Saussure, H. F.**, Monographie des Guêpes Solitaires ou de la tribu des Euméniens, etc. Paris, Masson, 1852, 6 parts, 286 pp., 22 pls.; **Schuster, L.**, *Eumenes maxillosa* de Geer. Zeitschr. wiss. Insektenbiol. 7, 1911, p. 27–28; **Southwick, E. B.**, The Parsnip Web-worm. Insect Life 5, 1893, p. 106–109 (*Eumenes fraterna*); **Strand, E.**, Ein nordamerikanisches Eumenidennest nebst deskriptiven Bemerkungen über die zugehörigen Wespen. Ent. Mitt. 3, 1914, p. 116–118, 1 fig. (*Odynerus birenimaculatus*); **Tandy, M.**, The Carpenter Mud Wasp (*Monobia quadridens*). Ent. News 19, 1908, p. 231–232; **Taylor, L. H.**, Notes on the Biology of Certain Wasps of the genus Ancistrocerus (Eumenidæ). Psyche, 29, 1922, p. 48–65, 1 pl.; **Turner, C. H.**, A Week with a Mining Eumenid: An Ecologico-Behavior Study of the Nesting Habits of *Odynerus dorsalis* Fabr. Biol. Bull. 42, 1922, p. 153–172, 6 figs.; **Westwood, J. O.**, Notice of the Habits of *Odynerus antilope*. Trans. Ent. Soc. London 1, 1835, p. 78–80; **Williams, F. X.**, Notes on the Habits of Some Wasps that Occur in Kansas, with Description of New Species. Kans. Univ. Sc. Bull. 8, 1913, p. 223–230, 1 pl.; Philip. Wasp Studies, *l.c.* p. 149–

164 (Eumeninæ); **Xambeu, V.,** Nidification des Euménides. Naturaliste 29, 1910, p. 57–58.

8. For the purpose of facilitating reference the following list of selected literature on the social wasps is divided into sections:

VESPIDAE (General). **André, Ed.,** Les Guêpes, in André, Ed., Spécies des Hyménoptères d'Europe et d'Algérie. 2, 1883, p. 405; **von Dalla Torre, K. W.,** Fam. Vespidæ in Wytsman's Genera Insectorum fasc. 18, 1904; **Bequaert, J.,** A Revision of the Vespidæ of the Belgian Congo, etc. *l.c.*, 1918; **Ducke, A.,** Ueber Phylogenie und Klassifikation der sozialen Vespiden. Zool. Jahrb. Syst. 36, 1914, p. 303–330, 17 figs.; **Möbius, K. A.,** Die Nester der geselligen Wespen. Beschreibungen neuer Nester, etc. Abhandl. naturw. Ver. Hamburg 3, 1856, p. 121–171, 19 pls.; Vergleichende Betrachtungen über die Nester der geselligen Wespen. Wiegmann's Archiv 22, 1856, p. 321–332, 1 pl.; **Pérez, J.,** Notes sur les Vespides. Act. Soc. Linn. Bordeaux 64, 1910, p. 1–20; **Roubaud, E.,** Recherches Biologiques sur les Guêpes Solitaires et Sociales d'Afrique. Ann. Sc. Nat. Zool. (10) 1, 1916, p. 1–160, 34 figs.; **Rudow, F.,** Einige Kunstbauten von Faltenwespen. Ill. Wochenschr. Ent. 2, 1897, p. 321–326; Beschreibung einiger ausländischer Wespennester. Ent. Zeitschr. Guben 20, 1906, p. 185–187, 201–203, 12 figs.; Das Leben der Falten-Wespen, Vespidæ. Ent. Rundschau 30, 1913, p. 67–69, 74–76, 81–82, 88–90, 100–102, 112–114, 118, 125–126, 31 figs.; **de Saussure, H. F.,** Études sur la famille des Vespides. Monographie des Guêpes Sociales ou de la tribu des Vespiens. Paris, Masson, 1853, 256 pp., 37 pls. Suite 1856–1857, 144 pp., 3 pls.; Suite, Partie 3. Monographie des Masariens, *ibid.* 1856, 352 pp., 16 pls.

VESPIDAE (Neotropical). **Brèthes, J.,** Sur quelques nids de Vespides. Anal. Mus. Nac. Buenos Aires (3) 7, 1902, p. 413–418, 1 pl.; Contribución al estudio de los Vespidos sudamericanos y especialmente argentinos. *ibid.* (3) 2, 1903, p. 15–39; **du Buysson, R.,** Nid de la *Polybia phthisica* Fabr. Bull. Soc. Ent. France 1899, p. 129–130, 1 fig. Monographie des Vespides du genre Nectarina. *ibid.* 74, 1905, p. 537–566, 6 pls.; Monographie des Vespides appartenant aux genres Apoica et Synœca. *ibid.* 75, 1906, p. 333–362, 7 pls.; **Ducke, A.,** Sobre as Vespidas sociaes do Pará. Bol. Mus. Goeldi Pará 4, 1904, p. 317–374, 4 figs.; 1. Supplemento. *ibid.* 4, p. 652–698, 1 fig.; Nouvelles Contributions à la Connaissance des Vespides Sociales de l'Amérique du Sud. Rev. d'Ent. 1905; Novas contribuicoes para o conhecimento das vespas (Vespidæ sociales) da regiao neotropical. Bull. Mus. Goeldi Pará 5, 1908, p. 152–200; Revision des Guêpes Sociales Polygames d'Amérique. Ann. Hist. Nat. Mus. Nat. Hungar. 8, 1910, p. 449–544; **Edwards, H. M.,**

Note sur le nid de l'*Epipona tatua* (*Polistes morio* F.). Ann Soc. Ent. France 1, 1843, Bull. p. 24–25; **Gallardo, A.**, Algunas observaciones biologicas sobre los camuaties. Rev. Jard. Zool. Buenos Aires 4, 1908, p. 21–23; **von Ihering, H.**, Zur Biologie der sozialen Wespen Brasiliens. Zool. Anzeig. 19, 1896, p. 449–453; L'état des Guêpes sociales du Brésil. Bull. Soc. Zool. France 21, 1896, p. 159–162; A Contribution to the Biology of the Social Wasps of Brazil. Transl. by E. E. Austen. Ann. Mag. Nat. Hist. (6) 19, 1897, p. 133–137; Hibernation of Social Wasps. Berlin. Ent. Zeitschr. 42, 1897, p. 139; **von Ihering, R.**, As Vespas sociaes do Brazil. Rev. Mus. Paulista 6, 1904, p. 97–315, 3 figs.; **Lassaigne, J. L.**, Examen chimique du miel de la Guêpe Lechéguana. Mém. Mus. d'Hist. Nat. 2, 1824, p. 319–320; **Latreille, P. A.**, Notice sur un insecte hyménoptère de la famille de Diploptères, connu dans quelques parties du Brésil sous le nom de Guêpe lechéguana et récoltant du miel (*Polistes Lecheguana*). Mém. Mus. d'Hist. Nat. 11, 1824, p. 313–318 (fig. on Pl. 12 of Tome 12); **Lucas, M. H.**, Quelques Remarques sur les Nids des *Polybia scutellaris* et *liliacea*, Hyménoptères Sociaux de la tribu des Vespides. Ann. Soc. Ent. France (4) 7, 1867, p. 365– 370, 1 pl.; Nouvelle Espèce de Polybia. *ibid.* 1879, p. 363–372, 1 pl.; **Pellett, F. C.**, Nectarina in Texas. Science N. S., 55, 1922, p. 644–645; **de Saint-Hilaire, A. F. C. P.**, Relation d'un empoisonnement causé par le miel de la Guêpe Lechéguana. Mém. Mus. d'Hist. Nat. 12, 1825, p. 293–348, fig.; **von Schulthess-Rechberg, A.**, Parapolybia Saussure, Vespidæ sociales. Mitt. Schweiz. Ent. Ges. 12, 1913, p. 152–164, 2 pls.; **Schupp, P. A.**, Leben und Nest der Canguaxi (*Polybia scutellaris* (White) Sauss.). Natur. u. Offenbar. 42, 1896, p. 143–151; **Strand, E.**, Ueber das Nest einer neotropischen Wespe, *Polybia occidentalis* Ol. Ent. Mitteil. 3, 1914, p. 171–172, 1 pl.; **Wasmann, E.**, Beutetiere von *Polybia scutellaris* (White) Sauss. Zool. Anzeig. 20, 1897, p. 276– 279; **White, A.**, Description of *Myrapetra scutellaris*, a South American Wasp which collects Honey. Ann. Mag. Nat. Hist. 7, 1841, p. 315– 348; **Zavatteri, E.**, Une nouvelle guêpe sociale polygame du Brésil. Ann. Mus. Nat. Hungar. 9, 1911, p. 343–344 (*Synœcoides mocsaryi*).

Vᴇsᴘɪᴅᴀᴇ (Paleotropical). **Bequaert, J.**, A Revision of the Vespidæ of the Congo, etc., 1918, *l.c.;* **Bingham, C. T.**, (Nests of *Ischnogaster eximius* Bingh. and *nigrifrons* Smith). Journ. Bombay Nat. Hist. Soc. 5, 1890, p. 244–246; **du Buysson, R.**, Monographie des Vespides du genre Belonogaster. Ann. Soc. Ent. France 78, 1909, p. 199–270, 6 pls.; **Roubaud, E.**, Recherches Biologiques sur les Guêpes Solitaires et Sociales d'Afrique. 1916 *l.c.;* Evolution de l'instinct chez les Guêpes sociales d'Afrique du genre Belonogaster Sauss. C. R. Acad. Sc. Paris 151, 1910, p. 553–556; **Wagner, W.**, Nestbau von *Belonogaster junceus* Oliv.

Nerthus 4, 1902, p. 569–570, 1 fig.; **Williams, F. X.,** Philip. Wasp Studies. *l.c.,* p. 166–176 (Stenogaster).

POLISTES. **Adlerz, G.,** Utvecklingen af ett Polistes samhälle. Ent. Tidskr. 25, 1904, p. 97–106; **Bertrand, G.,** Examen du miel produit par un Poliste de Basse-Californie. Bull. Mus. d'Hist. Nat. 1895; **Brimley, C. S.,** Male *Polistes annularis* Survive the Winter. Ent. News 19, 1908, p. 107; **Enteman, Miss M. M.,** Some Observations on the Behavior of the Social Wasps. Pop. Sc. Month. 61, 1902, p. 339–351; **Girault, A. A.,** Notes on the Predaceous Habit of *Polistes rubiginosus* St. Farg. Canad. Ent. 39, 1907, p. 355–356; Incidental Observations on a Queen *Polistes pallipes* Lep. St. Farg. etc. Bull. Wis. Nat. Hist. Soc. 9, 1911, p. 49–63; **Guignon, J.,** *Polistes gallica* et son nid. Feuille jeune Natural. (5) 42, 1912, p. 117; **Hingston, R. W. G.,** A Naturalist in Himalaya. Boston, Small, Maynard & Co., 1922, p. 180 *(Polistes hebræus)*; **Lubbock, Sir J.,** Ants, Bees and Wasps. Rev. ed. New York, Appleton, 1894 (Vespa, Chapt. 11.); **Marchal, P.,** Observations sur les Polistes. Bull. Soc. Zool. France 21, 1896, p. 15; Arrivé, coucher et départ des Martinets, en 1901, au Creusot. Bull. Soc. Nat. Autun. No. 14, 1901, Proc. Verb., p. 240–241; Sur le mode de nidification des Polistes, *ibid.,* p. 241–242; **Rudow, F.,** Zur Ueberwinterung von Polistes. Insectenbörse 15, 1898, p. 33; **Ferton, C.,** Notes détach. *l.c.* 1, p. 128–129, 4, p. 572–573 *(Polistes gallica)*; **Rau, P.,** and **N.,** Wasp Studies. *l.c.,* Chapt. 12 *(P. pallipes, rubiginosus, annularis* and *bellicosus)*; **Schulz, W. A.,** Das Nest von *Polistes hebræus* F. Verh. zool. bot. Ges. Wien 55, 1905, p. 490–493; **Westwood, J. O.,** On the Proceedings of a Colony of *Polistes gallica* introduced into my Garden at Hammersmith from the Neighborhood of Paris. Trans. Ent. Soc. London 4, 1845, p. 123; **Wheeler, W. M.,** Vestigial Instincts in Insects and Other Animals. Amer. Journ. Psychol. 19, 1908, p. 1–13 (honey stored by Polistes).

VESPA. **von Buttel-Reepen, H.,** Zur Biologie von *Vespa germanica.* Psychobiol. u. biol. Beobacht. Naturw. Wochenschr. 6, 1907, *l.c.;* (Jacobson's observation on *Vespa analis* Fabr.) *ibid.,* aiso Auffällige Eiablage vei *Vespa media,* and Eine absonderliche Nestmodifikation von *Vespa vulgaris. ibid.;* **du Buysson, R.,** Monographie des Guêpes ou Vespa. Ann. Soc. Ent. France 72, 1903, p. 260–288; 73, 1905, pp. 485–555, 566–634; **Fabre, J. H.,** Souvenirs Ent. *l.c.* 8, Chapt. 19–21; **Forel, A.,** Quelques Observations biologiques sur les guêpes. Bull. Soc. Vaud Sc. Nat. (4) 31, 1896, p. 312–314; **Frohawk, F. W.,** Attitude of Hybernating Wasp. Entom. 36, 1903, p. 33–34, 2 figs.; **Fyles, T. W.,** The Paper-Making Wasps of the Province of Quebec. 33rd Ann. Rep. Ent. Soc. Ontario 1903, p. 69–74, 5 figs.; **Hess, W.,** Neue Beobachtungen über das Leben der Hornisse. Ent. Nachr. 11, 1885, p. 218; **Hingston, R. W. G.,**

A Naturalist in Himalaya. *l.c.* p. 177 *et seq.* (*Vespa orientalis* and *magnifica*); **Howes, P. G.**, The Original Paper Makers. The Paper Wasp and Its Life from Egg to Death. Scient. Amer. 105, 1911, p. 102, 109, 10 figs.; **Imhoff, L.**, Lebensweise der gemeinen Wespe. Ber. Verh. naturf. Ges. Basel 8, 1849, p. 41; **Janet, C.**, Sur *Vespa media, V. silvestris* et *V. saxonica.* Mém. Soc. Acad. l'Oise 16, 1895, p. 28, 9 figs.; Sur *Vespa germanica* et *V. vulgaris*, 26 pp. 5 figs.; Sur *Vespa crabro* L. Histoire d'un Nid depuis son Origine. Mém. Soc. Zool. France 8, 1895, 140 pp., 41 figs.; Observations sur les Guêpes. Paris, C. Naud., 85 pp., 30 figs.; **König, C.**, Was wussten die alten Griechen und Römer von den Wespen und Hornissen? Ill. Wochenschr. Ent. 1, 1896, p. 261–266; **Kristof, L. J.**, Ueber einheimische, geselliglebende Wespen und ihren Nestbau. Mitt. naturw. Ver. Steiermark, 1878, p. 38; **Latreille, P. A.**, Observations sur quelques Guêpes. Ann. Mus. d'Hist. Nat. 1, 1802, p. 287–294, 1 pl.; **MacGarvie, J.**, Observation on the large brown Hornet of New South Wales, with reference to instinct. Edinb. New Phil. Journ. 4, 1828, p. 237–242; **von Malinowsky**, Beiträge zur Naturgeschichte der *Vespa crabro.* Magaz. Berlin. Ges. naturf. Freunde 2, 1808, p. 151–156; **Marchal, P.**, La Reproduction et l'Evolution des Guêpes sociales. Arch. Zool. Expér. Gén. (3) 4, 1896, p. 1–100, 7 figs.; **Marlatt, C. L.**, Observations on the Habits of Vespas. Proc. Ent. Soc. Wash., 1900, p. 79–84; **Morley, M.**, Wasps and Their Ways. New York, Dodd, Mead & Co. 1900; 1902, **Mosely, S. L.**, Tree Wasps and Their Nests. Halifax Natural. 7, 1902, p. 111–113, 1 fig.; **Müller, P. W. J.**, Beiträge zur Naturgeschichte der grossen Hornisse, *Vespa Crabro* F., einige an einem gezähmten Hornissen-Neste angestellte Beobachtungen enthaltend. Germar. Mag. Ent. 3, 1818, p. 56–68; **Murray, A.**, Wasps: Their Life-History and Habits. Trans. Edinb. Field Nat. Micr. Soc. 3, 1898, p. 342–352; **Newport, G.**, On the Predaceous Habits of the Common Wasp, *Vespa vulgaris* Linn. Trans. Ent. Soc. London 1, 1836, p. 228–229; **Ormerod, E. L.**, Contributions to the Natural History of the British Vespinæ. Zoologist 17, 1859, p. 6641–6655; **Pack-Beresford, D. R.**, The Nesting Habits of *Vespa rufa.* Irish Natural. 11, 1902, p. 94–95; Notes on Wasps. *ibid.* 12, 1903, p. 15–17; **Pechlander, E.**, Zum Nestbau der *Vespa germanica.* Verh. Zool.-bot. Ges. Wien 54, 1904, p. 77–79; **Rau, P.** and **N.**, Wasp Studies, *l.c.*, Chapt. 12 (*Vespa germanica* and *V. maculata*); **Rudow, F.**, Triumph der Züchtung, Insektenbörse 15, 1898, p. 74; **Sahlberg, J.**, Bo af *Vespa rufa.* Medd. Soc. F. et Fl. Fenn. 26, 1900, p. 44; **Schenck**, Die deutschen Vesparien. Wiesbaden 1861; **Shuckard, W. E.**, On the Pensile Nests of British Wasps. Mag. Nat. Hist. 3, 1839, p. 458–460; **Smith, F.**, Observations on the Habits of *Vespa norwegica* and *Vespa germanica.*

Zoologist 10, 1852, p. 3699–3703, 1 fig.; On the manner in which *Vespa rufa* builds its Nests. *ibid.* 14, 1856, p. 5169–5174; **Walker, F. A.**, Hornets: British and Foreign. Journ. Trans. Victoria Inst. London 33, 1901, p. 362–392; **Williams, F. X.**, Philip. Wasp Studies *l.c.*, p. 164–166 (*Vespa luctuosa* and *deusta*).

VESPA (Parasitic species). **Bequaert, J.**, On the Occurrence of *Vespa austriaca* Panzer in the Northeastern United States. Bull. Brooklyn Ent. Soc. 11, 1916, p. 101–107; Bees and Wasps. Scientific Results of the Katmai Expedition of the National Geographic Society, Ohio Journ. Sci. 20, 1920, p. 292–297; **Carpenter, G. H.**, and **Pack-Beresford, D. R.**, The Relationship of *Vespa austriaca* to *Vespa rufa.* Ent. Month. Mag. (2) 14, 1903, p. 230–242, 1 pl.; The Relationship of *Vespa austriaca* to *Vespa rufa.* Irish Natural. 12, 1903, p. 221–238, 1 pl., 2 figs; **Cuthbert, H. K. G.**, A Mysterious Irish Wasp, *Vespa austriaca* Panz. (*arborea* Smith). *ibid.* 6, 1897, p. 285–287, 1 pl.; **Fletcher, J.**, *Vespa borealis* an Inquiline? Ann. Ent. Soc. Amer. 1, 1908, p. 30; Psyche 15, 1908, p. 16; **Morawitz, F.**, Ueber *Vespa austriaca* Panz. und drei neue Bienen. Bull. Soc. Imp. Nat. Moscow 37, 1864, p. 439–449; **Pack-Beresford, D. R.**, Another Nest of *Vespa rufa-austriaca.* Irish Natur. 13, 1904, p. 242–243; **Robson, J. E.**, *Vespa austriaca*, a Cuckoo Wasp. Science Gossip, n.s. 5, 1898, p. 69–73; **Schmiedeknecht, O.**, Ueber einige deutsche Vespa-Arten. Ent. Nachr. 7, 1881, p. 313–318; **Sladen, F. W. L.**, The Genus Vespa in Canada. Ottawa Natural. 32, 1918, p. 71–72; **Sharp, D.**, *Vespa rufa + austriaca.* Ent. Month. Mag. (2) 14, 1903, p. 297–299; **Wheeler, W. M.**, and **Taylor, L. H.**, *Vespa arctica* Rohwer, a Parasite of *Vespa diabolica* De Saussure. Psyche 28, 1921, p. 135–144, 3 figs.

8. The storing of termites in its nest by Polybia seems first to have been observed by H. von Ihering (Zur Biologie der socialen Wespen Brasiliens *l.c.*, 1896, p. 451). The following year Wasmann (Beutetiere von *Polybia scutellaris*, etc., *l.c.*, 1897) published more detailed observations made by Father Schupp in Rio Grande do Sul. He saw hundreds of termite-laden wasps simultaneously entering a nest. All the specimens carried in were males and females of *Eutermes ater* Hagen, and had evidently been captured during their dispersion flight. The wasps had also brought in males of three species of ants: *Dorymyrmex pyramicus* (about 100), *Wasmannia* sp. (2) and *Ponera trigona* ? (1).

July 16, 1920, I found at Kartabo, British Guiana, a small and very delicate nest of a wasp which was later identified by Dr. J. Bequaert as belonging to *Protopolybia sedula* Sauss. var. *palmarum* Blanch., but which Dr. J. C. Bradley believes to be *P. occidentalis* Oliv. This nest was attached to the lower surface of a leaf and was only about two inches in diameter. Its three small combs and the spaces between them and the involucre were packed with dead male and female termites, some of which still bore wings. The total supply would have filled a tablespoon. Dr. Bradley sends me the following notes on three *Protopolybia* nests which he and Harris encountered in Brazil. The first, which was found at Lassance, Minas Geraes, Nov. 13, 1919, belonged to *P. occidentalis*. Dr. Bradley writes: "I unfortunately broke the nest as it was quite soft and was amazed to find it literally packed with deälated termites, which had been flying in abundance for a few days. Evidently the wasps had been preparing a regular feast. About half a pint fell out into the cloth in which I had wrapped the nest and every cell and gallery seemed to be full of them." Four days later Harris captured another nest of wasps that had also stored termites. In this case also the wasp seems to have been *P. occidentalis*. A third nest, containing many termites was taken by Dr. Bradley Dec. 19, at Corumba, in the state of Minas Geraes. The wasp proved to be *P. occidentalis* var. *ruficeps* Schrottky. These observations indicate that termite storing is a regular industry of *P. occidentalis* and its varieties throughout tropical South America and that it enables these wasps to make good use of the local and occasional swarms of the white ants.

LECTURE III

1. For much interesting honey-bee lore the following works, especially those of Cowan, Glock and Robert-Tornow, may be consulted: **Billiard, R.,** Die Biene und die Bienenzucht im Altertume. Trans. by Breiden. Millingen, Th. Gödden, 1904, 108 pp., 25 figs.; **Cantipratanus Th.,** Liber apum, aut de apibus mysticis, sive de proprietatibus apum, etc. Strassburg, about 1472; Cologne 1475; Bonum universale de apibus, in quo mirifica Apum republica universa vitæ bene et christianæ instituendæ ratio traditur, etc. Duaci, Beller 1627. Earlier editions 1597, 1605 and 1624; **Cowan, F.,** Curious Facts in the History of Insects. Phila., Lippincott 1865; **Glock, J. P.,** Die Symbolik der Bienen und ihrer Producte in Sage, Dichtung, Kultur, Kunst und Bräuchen der Völker. 2nd edit. Heidelberg, Weiss'sche Univers.-Buchh., pp. XII + 411, 1897; **Koeppen,** Bienengeschichte der älteren Zeiten aus Sinnbildern, auf Münzen, Steinen, Gemälden, Abhandl. Arb. Oberlaus. Bienengesell. 3, 1768–1769, 83 pp.; **de Loche, Count F. M.,** De l'abeille chez les anciens; mémoire qui comprend l'historique de leur connaissances sur cet insecte, particulèrement à l'égard de sa génération, comparées aux decouverts

des modernes. Mém. Soc. Acad. Savoie 4, 1830, p. 208–235; **Magerstedt,** Die Bienenzucht der Völker des Alterthums, insbesondere der Römer. 1851, p. VI + 128; **Menzelius, W.,** Mythologische Forschungen und Sammlungen. Monographie der Biene 1842; **Robert-Tornow, W.,** De Apium Mellisque apud Veteros Significatione et Symbolica et Mythologia, Berolini, Apud Weidmannos 1893; **Schenck, C. F.,** Die Honigbiene vom Hymettus. Jahrb. Ver. Naturk. Herzogt. Nassau 14, 1859, p. 417–419; **Schenckling, S.,** Die Honigbiene in Sage und Geschichte. Insektenbörse 13, 1896, p. 186–187; **Theen, H.,** Die Biene im deutschen Volksglauben. Illustr. Wochenschr. Ent. 2, 1897, p. 530–534, 551–553, 563–567.

2. **Osten Sacken, C. R.,** On the So-called Bugonia. Bull. Soc. Ent. Ital. 25, 1893, p. 186–217; On the Oxen-born Bees of the Ancients (Bugonia) and their Relations to *Eristalis tenax.* Heidelberg, J. Hoerning. 1894, pp. XIV + 80.

In addition to the literature cited above, accounts of the bugonia may be found also in the following: **Camerarius, J. R.,** Apes ex bove; Apes cum bovum stercore delectuntur; Apium et vesparum ictum signa; Apes num coeant. Syllog. Memorab. 1654 (?), Cent. 17, part. 49–55, p. 1401; p. 1405; **Heikertinger, F.,** Die Bienenmimikry von Eristalis. Zeitschr. wiss. Insekbiol. 14, 1918, p. 1–5, 73–79; **Overbeck, J. A.,** Epistola de apibus im sancta scriptura male audientibus. Stadæ 1752; Gedanken von der Bugonia der Alten, da man sich getraute aus einem todten Rindvieh einen ganzen Schwarm lebendiger Bienen hervor zu bringen. Hamburg, Vermischte Bibl. 3 (about 1760); **Sterne, C.,** Die Beseitigung verbrauchter Lebenshüllen. 2. Das Märchen von der künstlichen Bienenerzeugung. Prometheus 6, 1895, p. 310–312, 325–328.

3. Among a lot of Baron Osten Sacken's unpublished notes on the " Bugonia," which came into my possession many years ago, I find two letters, one by the eminent French zoologist, Professor A. Giard, the other a reply by Osten Sacken. Apart from their historical value, these letters are so erudite and interesting that I have translated them for publication in this place. Giard's letter has a bearing on a casual observation which I made during the summer of 1920 at Kartabo, in British Guiana. A huge catfish, taken nearby in the Mazaruni River, was brought to the Tropical Laboratory and cut up on the shore. In less than half an hour swarms of large white butterflies (Papilio) and large bees (*Centris personata*) came to the fish's blood, which had spread over the rocks, and like the shades of the dead in the Odyssey, eagerly lapped it up. The Centris is also attracted by human perspiration and often annoyed us by flying about our heads while we were

boating on the rivers. The case of the meat-eating Trigonas cited at p. 124 is also interesting in connection with Giard's letter.[1]

Giard's Letter to Osten Sacken

14 RUE STANISLAS, PARIS
November 6, 1893.
DEAR AND HONORED COLLEAGUE:

On returning to Paris I found the very learned and very amusing memoir on Bugonia, which you kindly sent me and for which I cordially thank you. I read it with the greater interest, because this historico-biological problem has preoccupied me for many years. I had even reached what I believe to be a rational solution and had often referred to it in my courses, although I have published only a few lines on the subject. For some time I believed that I was the only zoologist who had seen anything more than a poetical myth in the old account of Aristæus, and not till two years ago did I learn that the matter had been discussed by another biologist, an entomologist, whose talent as an observer I admire but whose general ideas are so very reactionary (I speak from the scientist's point of view) that they make my flesh creep. I allude to J. H. Fabre of Avignon.

In the "Souvenirs Etomologiques," 4th series, 1891, p. 260, I find

[1] The following observation on *Vespa orientalis* recently published by R. W. G. Hingston (A Naturalist in Himalaya. Boston, Small, Maynard & Co., 1922), is also of interest in connection with the Giard-Osten Sacken correspondence: "Far and wide these wasps had scattered over the country. Wherever filth and refuse had accumulated there were the wasps to be seen searching every corner for a precious burden. Down the village street, exploring every nook and cranny in the foul bazars, boldly entering every shop, busy amongst the mules and camels of our transport and far around over the fertile fields, these industrious insects were engaged from morning to night in a continual search for plunder. Fragments of decomposing meat, decaying remnants of fish or anything of a sugary nature was enveloped in a swarm, torn into pieces by a hundred busy jaws and carried off to the nest. These wasps at certain times, are the natural scavengers of the country. With the kites, the pariah dogs and the dung-rolling beetles, they help to cleanse the village of its refuse. At one place they had crowded round a dead pigeon. Nothing was left of the flesh but a few tough fragments on the wings, which the wasps were unable to separate from the bases of the quills. They dragged about the feathers and the whole wings in the attempt to bite away the hard fibers, and one of them, unable to detach the flesh and unwilling to desert its provender, sailed away for the nest carrying in its tiny claws a large pinion about five inches in length."

the following passage, in which in his usual rather emphatic manner, Fabre expresses in regard to the Bugonia the very same opinion which you have so learnedly expounded: "The impetuous Philanthus, the strangler of bees, upset my preconceived ideas, when I placed it in the presence of the Eristalis (*E. tenax*), the Virgilian bee. He, the Philanthus, mistake this fly for a bee! Go to! The common man is deceived, antiquity was deceived, as the Georgics bear witness, which describe the birth of a swarm of bees from the carrion of a sacrificed ox, but the Philanthus is not deceived. To his senses, more clairvoyant than our own, the Eristalis has the odor of the Dipteron, the friend of filth, and nothing more."

Notwithstanding the very weighty arguments which you adduce in favor of this manner of viewing the matter, I believe that the ancients were by no means such poor observers, nor less sagacious than the Philanthus. It is said that the development of the individual repeats that of the race. Well, even as a child I never for one moment mistook the Eristalis for a bee. With its sombre livery this fly does, perhaps, resemble our black honey-bee, but it is very different from the blonde Ligurian bee (*Apis ligustica*) or the bee of Egypt (*A. fasciata*), the very races that occur in the countries where the Bugonia legend arose. How could the shepherd, who is able to distinguish the individual sheep in his flock, confuse two creatures of such different complexions? I believe, on the contrary, that the ancients made a very precise observation on the bee, but that its significance escaped them because of their rudimentary knowledge of physiology. Both bees and wasps sometimes seek out the flesh of recently slaughtered mammals, since they are attracted by their blood and viscera. This explains the rôle played by the entrails and particularly by the liver, that glycogenic organ *par excellence*, in the Bugonia question, and enables us to explain also the choice of an herbivore and of a young but adult animal. (The normal proportion of sugar in the blood, according to Claude Bernard, is as follows: man 0.90, ox 1.27, calf 0.93, horse 0.91, per 1000). Even the method of slaughter (without effusion of blood), recommended in the Georgics, is not without its significance. Castre, Bataillon, and others have, in fact, demonstrated that there is a post-mortem hyperglycæmia in animals killed by suffocation.

That bees will seek sugar in other places besides flowers is certain. It has been possible to keep swarms alive on a diet of artificial sugars. Not infrequently bees may be seen lapping up the honey-dew of plant-lice. Nor is it very unusual to see bees in the urinals of our large cities, a fact which did not escape Réaumur (Mémoirs V, p. 216). I am convinced that they are attracted to such places not so much by the humidity as by the sugar in the urine of the diabetics who are so numerous among our urban populations.

But there are additional arguments. Bees, like wasps, though perhaps less often, visit butcher-shops. I have called attention in a short note to the interest attaching to Réaumur's observations on this habit in wasps, a fact already indicated by the ancient Pliny. The observations may be easily repeated in villages where the butcher himself kills and exposes the very fresh meat. In cities the flies usually seen in abattoirs are Calliphora and Lucilia. It was a young lion that Samson killed, and it was on the following day (*post diem*), when the flesh was still fresh, that he saw the bees (the honey-comb was evidently interpolated by the oriental imagination). The commentators and S. Bochart translate *post diem* by a whole year, but this agrees neither with the text nor, in my opinion, with zoology.

No doubt, bees are rarely seen on decomposing animals, but the Bugonia is cited as an exception by all the authors, because it is admitted that there is among bees another method of generation like that of other insects. It is clear, furthermore, that this observatior would be more frequently made in southern countries, where the Bugonia question arose, since in their dry climates bees would find in the blood both sugar as nutriment and water to quench their thirst. Note also that it is especially in times of scarcity of flowers, that the phenomenon is described as occurring. Witness Aristæus: "*Amissis ut fama, apibus morboque fameque.*" But if I willingly concede that bees are only rarely and exceptionally seen on fresh meat, you will admit that one even more rarely and more exceptionally sees an Eristalis settling on a cadaver or encounters the larvæ of Eristalis in carrion. I, for one, have never seen either of these phenomena. My laboratory at Wimereux, whenever we dissect some large animal, is soon invaded by swarms of Calliphora, which arrive from all points of the compass, with a smaller number of Lucilia and Sarcophaga. We never see any Eristalis, although they frequent one of the outhouses in which the water-closets are situated and where *Homalomyia scalaris* Fal., *Scatopse notata* L., etc., abound.

You will cite as an objection the passage in Zetterstedt concerning *E. anthophorinus* (Diptera Scandinaviæ II, p. 666) where he speaks of Eristalis as flying to the cadaver of a sheep. But this cadaver was floating in a ditch and it was only occasionally that the flies, interrupting their oviposition, which occurred in flight ("*celerrime circumvolando*"), reposed for a moment on the emergent portions of the cadaver ("*in cadaveris parte supra aquam elevata interdum sedentia*"), just as they would have settled on the leaves of a floating plant or some other support. The Eristalis larvæ certainly live in very foul water, but they require an aquatic medium and do not quit it till they have completed their transformations.

This, in a few words, is the solution which I have conceived of this ancient problem. I am sending you a short note which I published sometime ago on wasps and glycogen. It contains what I have given in my courses and was suggested by an article that appeared in the "Naturaliste." This I also send you. The author, Dr. Lahille, cites some lines which did not appear in my note but in a work published in 1879, where the note was analyzed.[1]

I had hoped to finish this year and to be able to send you soon a note on the larvæ of Cecidomyia, which have long occupied me, but during the past six months my health has left much to be desired and I have been unable to complete the work.

Accept, dear and honored colleague, the assurance of my highest esteem.   GIARD.

Osten Sacken's Reply to Giard.

HEIDELBERG, November 19, 1893.

DEAR PROFESSOR GIARD:

Although your very interesting letter of November 5 reached me several days ago, writing an answer has consumed some time. It is indeed strange that we should have met on the same territory and I greatly regret that we are not marching in accord, but as we have the same goal, which is the search for truth, I hope that we may arrive together. As you are the stronger, I would ask nothing better than to be converted. I recall a verse which I found somewhere: " Show me my vanquisher and I will run to embrace him," or words to that effect. At any rate, it expresses my feelings.

I had busied myself with the Bugonia question for some time before 1886 (" Bugonia," p. 188) and had made a rather voluminous collection of notes when, in the spring of that year, the secretary of the Italian Entomological Society invited me to take part in the celebration of the society's twenty-fifth anniversary by contributing an article. The occasion seemed to me to be propitious for launching my " Bugonia," especially as the subject had so many points of contact with Italy. In

[1] The papers cited by Giard are the following: *A. Giard*, Comme quoi les quêpes ont decouvert la fonction glycogénique du foie longtemps avant M. C. Bernard. Bull. Scient. Hist. Litt. Dep. Nord 7, 1875, pp. 49–51, and *F. Lahille*, Le Rôle du Foie chez les Anciens. Allégorie d'Aristée. Le Naturaliste 13, 1891, pp. 198, 199.

my exposition I attempted to be clear and concise, in order that I might be understood by the general public. At the same time I omitted certain arguments and details in the hope that I might use them on a future occasion, or perhaps in a future edition of the work. I will make use of some of this material in replying to your objections.

You say that the Eristalis do not attack cadavers and you throw doubt on the observations of Zetterstedt. This is at bottom your principal objection. It is true that the Eristalis do not oviposit on fresh cadavers and you are right in maintaining that the latter are invaded first by the Lucilias, Calliphoras, etc. But a carcass, after exposure to the air for some time, reaches a second stage of decomposition, during which a pool of fetid water forms beneath it. This is the moment when Eristalis lays her eggs, and it was this that Zetterstedt observed. You say: "But this cadaver was floating *in a ditch* and it was only occasionally that the Eristalis interrupting their *oviposition, which occurred in flight,* reposed for a moment on the emergent portion of the cadaver, just as they would have settled on the leaves of a floating plant or some other support, etc." I reply that it was not a *ditch,* that Zetterstadt saw, but a pool of fetid water formed beneath the cadaver. *The oviposition of the Eristalis did not take place during flight;* on the contrary, they alighted on the dry surface of the animal, a little above the liquid, and when the larvæ hatched they fell into the bottom of the puddle. Compare the description of the oviposition of Eristalis in Réaumur (Mémoirs IV, p. 475), which I regard as one of his *chefs-d'œuvres* of patient observation. Read the passage which begins with the words: "I have often had the pleasure of following with my eyes one of the bee-like flies that came flying about a basin, etc." You will notice a striking resemblance to Zetterstedt's description, you will find in it the "*circumvolando*" ("it hovers while describing divers circles") and you will notice that "it alights on the inner surface of the basin, a few inches above the surface of the water. There it will be seen to elongate its posterior end considerably, etc." Read the entire passage, which seems to me to be a marvel of description. Such a coincidence is not fortuitous; it proves the identity of the object observed by Zetterstedt and Réaumur.

I would call your attention to another agreement which I believe to be very significant, this time between Réaumur and the ancients. The rat-tailed larvæ of Réaumur comprise three species: *Eristalis tenax,* which he calls "la mouche à forme d'abeille" (bee-like fly), Helophilus, which he calls "la mouche à forme de guêpe" (wasp-like fly) and a third, smaller species, which is probably *E. arbustorum* (Réaumur, Mémoires IV, p. 440, 474). Among the ancients it is always a question of *the bee and the wasp* in the cadavers of the ox and horse. Compare the two passages of Ælian (II, 57, and I, 28); Ovid's Metamorphoses (XV, 365–368), his Fasti I, 376, the passage of Plutarch ("Bugonia" p. 205) and the citations from Galen given in my "Bugonia" (pp. 186, 206, 209). Thus, on the one hand, we have Réaumur, who observed the transformations of the rat-tailed larvæ and saw the "mouche à forme d'abeille" emerge from them and also another fly "en forme de guêpe," and on the other hand, the belief of the ancients that both bees and wasps develop in animal cadavers. How can we explain these two agreements and at the same time the further coincidence of the two pairs of insects? Is Réaumur's pair (Eristalis and Helophilus) identical with the pair of the ancients? Or does this latter pair represent a true bee and a true wasp? In other words, are we dealing with a case of identity, or a case of mimicry? It seems to me that the identity hypothesis is infinitely more probable than that of mimicry.

Having ascertained that the Eristalis oviposit on carcasses in an advanced stage of decomposition, we can hardly resist the conclusion that in ancient times, when decomposing carcasses were lying about in greater profusion than they are today, it was quite natural that a superstition like the Bugonia should have arisen. Réaumur, so far as we know, never saw the rat-tailed larvæ near animal carcasses, so that Zetterstedt's observation is, for the present, unique. We might supply the deficiency by making the experiment, but we must not lose sight of the fact that such an experiment to be conclusive would have to succeed. We could never be sure that we had fulfilled all the conditions required by Nature. It would be better to seek cases in places remote from civilization, in some isolated valley in Italy, in Turkey or in Asia. Who knows whether the superstition may not survive in such localities even at the present day? The passage which I found in Massoudi proves that it was current in Arabia in the tenth century of our era (Bugonia, p. 201).

Your hypothesis does not account for the origin of the superstition and the constant simultaneous mention in ancient literature of bees, wasps and cadavers. You seem even to deny the existence of the Bugonia of the ancients. If the thing had not existed, there would have been no such word. J. H. Voss, the commentator of the Georgics (edit. 1829, p. 227) says: "The Greeks, Carthaginians and Romans speak of it as of an every-day truth ("als von einer alltäglichen Wahrheit"). The Greeks called the bees quite simply "bugeneis" and "taurigenæ." Had the ancients used the cadavers solely for the purpose of attracting the bees, they would soon have noticed that the insects did not emerge from the cadavers, but came to them from the outside. The two descriptions which we possess of the Bugonia in Virgil and in the Geoponica expressly mention the larvæ which are seen to emerge from the filth. M. Lahille, when citing Virgil, omits the principal passage (Georgics IV., 309, 310): *Visenda modis animalia miris trunca pedum.* This passage, which was incomprehensible to the commentators, can only refer to the larva: " Then one sees marvellous animals, without legs " . . . which later transform themselves into winged insects.

The mention of the larvæ is still more explicit in the recital of Florentinus: " When the stable is opened little white animals are seen, resembling one another, immobile, developing wings, taking on the color of bees; the wings are short at first, trembling, unsuited for flight, the limbs feeble; desirous of the light, they take wing and strike against the windows, jostle one another," etc. See Aldrovandus, de Anim. Insectis, Bologna, 1602, p. 59 *supra.* The mention of the king, " who is born from the brain," and around whom the putative bees cluster, is evidently a superfetation caused by the resemblance of the Eristalis to the bees. I have reflected much on the process related by Florentinus and believe that I am able to furnish a rather plausible explanation of the various manipulations to which it has been subjected, but I will not bore you with the matter at present.

You say: " With its sombre livery the Eristalis probably resembles our black bee, but it is very different from the blonde Ligurian and the bee of Egypt," etc. As opposed to this I would cite the passage (Bugonia, p. 209): " The coloring of the abdomen of the honey-bees is variable; some varieties have very distinct brownish-yellow cross-bands at its base; just the same varieties occur in the colouring of the fly *E. tenax.*" I have before me 17 specimens of *E. tenax*, captured in different parts of Germany; the majority are of the blonde variety.

I have still to mention your good opinion of the ancients as observers of nature. This is, of course, merely a matter of appreciation, and I believe that I have given in my work a considerable number of examples of lack of discernment on the part of both ancients and moderns. When a scientific coryphæus like Galen believes in the Bugonia, what may we not expect of his predecessors? " It is difficult to explain how the worms are generated in plants, or the bees and wasps are generated in oxen and horses. These are spontaneous operations of Nature, just as no one has taught the birds to fly, nor us to hear, see or understand " (Galen in his work: An animal id sit, quod in utero contineatur, Chapter V, cited by Aldrovandus *loc. cit.*). This is a fine specimen of facile natural history! I believe that Saint Routine reigned among the ancients, just as he does among ourselves, that it was as easy for them as for us, to *look*, but quite as difficult for them as for us, to *see*.

I confess that the difficulty which has, perhaps, caused me the greatest embarrassment in explaining the Bugonia is to understand how men could busy themselves so long with the artificial production of bees and not perceive that the product was flies, which are never provided with stings and could not therefore be true bees. Must we content ourselves in this case with the sentence of Ecclesiastes: " *stultorum infinitus est numerus* " ? Or may we here introduce your idea as a subsidiary hypothesis, that the true bees arrived from without and mingled with the Eristalis during the operation, contributing thus to prolong the illusion of the servitors of Juba?

You see that I have taken considerable pains to convert you. Have I succeeded? I hope at least that you will not take it amiss that I have defended my thesis " Without Eristalis no Bugonia! " with some ardor. When one has long held an opinion it is often difficult to abandon it suddenly; to use one of Réaumur's expressions: " everything may become an obstacle to the human mind."

This is certainly a long letter. I had many things to tell you about the Cecidomyids, but I prefer to await the work which you promised to send me. I am very sorry to hear that you are ill; as for myself, I feel, after completing 65 years, that I shall soon be ripe — for the Eristalis!

C. R. OSTEN SACKEN.

4. The following selections from a voluminous bibliography of solitary bees are divided into three sections and cover a wide range of species:

SOLITARY BEES (General): **André, E.**, Inwiefern werden Insekten durch Farbe und Duft der Blumen angezogen? Beih. bot. Centralbl. 15, 1903, p. 427–470; **Brunelli, G.**, Collezionismo e ibernazione nell' origine degli istinti delle Api solitarie e sociali. Riv. Ital. Sc. Nat. 24, 1904, p. 60–64; **von Buttel-Reepen, H.**, Die stammesgeschichtliche Entsehung des Bienenstaates, sowie Beiträge zur Lebensweise der solitären und sozialen Bienen (Hummeln, Meliponinen, etc.). Leipzig, G. Thieme, 1903, XII + 138 pp., 20 figs.; Die phylogenetische Entstehung des Bienenstaates, sowie Mitteilungen zur Biologie der solitären und sozialen Apiden. Biol. Centralbl. 23, 1903, p. 4–31, 89–108, 129–154, 183–195, 19 figs.; Psychobiologische und biologische Beobachtungen an Ameisen, Bienen und Wespen. Naturw. Wochenschr.

n.f. 6, 1907; **Christ, J. L.**, Naturgeschichte, Classification und Nomenclatur der Insecten vom Bienen, Wespen und Ameisenge-schlecht, etc. Frankf. a.M., Hermann, 1791, 535 pp., 60 pls.; **Friese, H.**, Beiträge zur Biologie der solitären Blumenwespen (Apidæ). Zool. Jahrb. Syst. 5, 1891, p. 751–860, 1 pl.; Die Bienen Europas (Apidæ Europææ) nach ihren Gattungen, Arten und Varietäten auf ver-gleichend-biologischer Grundlage I–VI. Innsbruck-Berlin. 1895–1901; **von Ihering, R.**, Zur Frage nach dem Ursprung der Staatenbildung bei den sozialen Hymenopteren. Zool. Anzeig. 27, 1903, p. 113–118; **Kirby, W.**, Monographia Apum Angliæ. Ipswich, White. 2 vols., 1802; **Latreille, P. A.**, Nouvelles observations sur la manière dont plusieurs Insectes de l'ordre des Hyménoptères pourvoient à la subsistence de leur postérité. Ann. Mus. d'Hist. Nat. 14, 1809, p. 412–425, 1 pl.; Considérations nouvelles et générales sur les insectes vivant en so-ciété. *ibid.* 3, 1817, p. 39–410; **Lovell, J. H.**, The Origin of the Oligo-tropic Habit among Bees. Ent. News 24, 1913, p. 104–112; **Müller, H.**, Anwendung der Darwinschen Lehre auf Bienen. Verh. Naturk. Ver. Preuss. Rheinl. u. Westph. 1871, p. 1–96, 2 pls.; Die Befruchtung der Blumen durch Insekten und die gegenseitigen Anpassungen beider. Leipzig 1873; Alpenblumen, ihre Befruchtung durch Insekten und ihre Anpassungen an dieselben. Die Entwickelung der Blumenthätigkeit der Insekten. Kosmos 9, 1881; **Müller, W. H.**, Proterandrie der Bienen. Inaug. Dissert. Jena 1882, 45 pp.; **Robertson, C.**, Oligotropic Bees, Ent. News 23, 1912, p. 457–460; **Schmiedeknecht, O.**, Apidæ Europææ (Die Bienen Europas) 2 vols. Berlin, Gumperda 1882–1886; Die Hymenopteren Mitteleuropas. Jena, G. Fischer, 1907; **Verhoeff, C.**, Neue und bekannte Gesetze aus der Hymenopteren-Biologie. Zool. Anzeig. 15, 1892, p. 362; Beiträge zur Biologie der Hymenopteren. Zool. Jahrb. Syst. 6, 1892, p. 732; **Vogt, C.**, Untersuchungen über Thierstaaten. Frankf. Litterar. Anst. 1851, pp. XVI + 248, 3 pls.

SOLITARY BEES (Nonparasitic). **Adlerz, G.**, Om cellbyggnad och tjufbin hos *Trachusa serratulæ* Panz. Ent. Tidskr. 25, 1904, p. 121–129; **Alfken, J. D.**, *Halictus tumulorum* L. und seine Verwandten. Ent. Nachr. 25, 1899, p. 114–126; Ueber Leben und Entwicklung von *Eucera difficilis* (Duf.) Per. *ibid.* 26, 1900, p. 157–159; Die Bienen-fauna von Bremen. Abh. naturw. Ver. Bremen 22, 1913, 220 pp.; **Armbruster, L.**, Chromosomenverhältnisse bei der Spermatogenese soli-tärer Apiden (*Osmia cornuta* Latr.). Beiträge zur Geschlechts-bestimmungsfrage und zum Reduktionsproblem. Arch. Zellforsch. 11, 1913, p. 242–326, 3 pls., 10 figs.; Zur Phylogenie der Geschlechts-bestimmungsweise bei Bienen. Zool, Jahrb, Syst. 40, 1916, p. 328–388, 1 pl., 6 figs. (Halictus, Sphecodes); **Aurivillius, C.**, Ueber Zwischen-

formen zwischen socialen und solitären Bienen. Festschr. Lilljeborg
1896, p. 67–77 (*Halictus longulus*); **Bellevoye, A.**, Sur quelques
Abeilles maçonnes. Bull. Soc. Sc. Nat. Reims, Trav. 6, 1897, p. 111–
113 (Chalicodoma); **Bertoni, A. W.**, and **Schrottky, C.**, Die Nestanlage
von *Xylocopa frontalis* Oliv. Ent. Nachr. 26, 1909, p. 40–42;
du Buysson, R., Nidification de quelques Mégachiles. Ann. Soc. Ent.
France 71, 1903, p. 751–755; **von Buttel-Reepen, H.**, Orientierungs-
vermögen der *Osmia papaveris* Ltr. in Psychobiol. u. biol. Beob. *l.c.*;
Das gekrümmte Ei von *Halictus quadricinctus* F. *ibid;* Ueber die
Schlafstellung von Wespen und Bienen. *ibid.* Abnorme Tapezierkunst
einiger *Osmia papaveris.* Instincktsänderung? *ibid.;* Ueber die
jungfraüliche Zeugung (Parthenogenesis) bei einsamlebenden Bienen
und die Staatenbildung. Die "neue" Goeldische Parthenogenesis-
Hypothese. Bienenwirtsch. Centralbl. 1917, Nos. 1–5, 1917, 7 pp.;
Cockerell, T. D. A., Habits of *Spinoliella zebrata* (Cresson). Ent.
News 1915, p. 366; **Davidson, A.**, On the Nesting Habits of *Anthidium
consimile. ibid.* 7, 1896, p. 22–26; **Ducke, A.**, Die Bienengattung
Osmia Panz. Ber. nat.-med. Ver. Innsbruck 25, 1900, p. 1–323,
59 figs.; Biologische Notizen über einige südamerikanische Hymenoptera. Zeitschr. wiss. Insektenbiol. 1, 1905, p. 175–177 (Euglossa); **Dufour, L.**, and **Perris, E.**, Sur les Insectes hyménoptères
qui nichent dans l'intérieur des tiges sèches de la Ronce. Ann.
Soc. Ent. France 9, 1840, p. 1–32, 2 pls.; **Eversmann, E.**, Die Brut-
stellen des *Hylœus quadricinctus* Fabr. Bull. Moscow 19, 1846, p. 188–
193, 1 pl.; **Fabre, J. H.**, Études sur les Mœurs et la Parthénogenèse
des Halictes. Ann. Sc. Nat. Zool. (6) 9, 1879–80, p. 1–27, 1 pl.;
Études sur la Répartition des Sexes chez les Hyménoptères. *ibid.*
1884, p. 1–53; Souvenirs Ent. *l.c.*, 1. Chapt. 20–22, 2, Chapt. 7, 3.
Chapt. 7–10, 19 (Chalicodoma); 3. Chapt. 16–20, 4, Chapt. 6 (Osmia);
4, Chapt. 8–9 (Anthidium); 4, Chapt. 7 (Megachile) 8, Chapt. 7–9
(Halictus); Transl. by A. T. de Mattos as "The Mason Bees" and
"Bramble-bees and Others." New York, Dodd, Mead & Co., 1914 and
1915; **Ferton, C.**, Recherches sur les Mœurs de quelques espèces algé-
riennes d'Hyménoptères du genre Osmia Panz. Act. Soc. Linn. Bordeaux 44, 1890, p. 201; Seconde Note sur les mœurs de quelques
Hyménoptères principalement du genre Osmia Panzer de la Provence
ibid. 47, 1894, p. 203; Nouvelles Observations sur l'instinct des Hyménoptères gastrilégides de la Provence. *ibid.* 48, 1895, p. 241; Nouvelles
observations sur l'instinct des Hyménoptères gastrilégides de France
et de Corse. *ibid.* 52, 1897, p. 36–50, 1 pl.; Remarques sur les
Mœurs de quelques espèces de Prosopis Fabr. Bull. Soc. Ent.
France 1897, p. 58–61; Notes détach. *l.c.* 1901–1921 *passim* (many

observations, mainly on gastrilegous bees of the genera Osmia, Megachile, Anthidium, but also on Prosopis, Nomia, Halictus, Anthophora, etc.); J. H. Fabre Entomologiste, Rev. Scient. Sept. 16–23, 1916; **Friese, H.**, Der Nestbau von *Osmia bicolor* Sauss. Ent. Nachr. 23, 1897, p. 113–116, 1 fig.; Ueber Osmien-Nester. Ill. Zeitschr. Ent. 3, 1898, p. 193; Eine neue Nestanlage von *Anthidium lituratum* Ps. *ibid.* 4, 1899, p. 116, 2 figs.; Ueber eine Koloniebildung bei der Mörtelbiene (*Chalicodoma muraria* Retz). Allg. Zeitschr. Ent. 8, 1903, p. 313–315, 1 fig.; Ein Bienennest mit Vorratskammern (*Lithurgus dentipes* Sm.). Zeitschr. wiss. Insektenbiol. 1, 1905, p. 118–119; **von Frisch, K.**, Beitrag zur Kenntnis sozialer Instinckte bei solitären Bienen. Biol. Centralbl. 38, 1918, p. 183–188, 1 fig.; **Frison, T. H.**, Notes on the Life History, Parasites and Inquiline Associates of *Anthophora abrupta* Say, with some Comparisons with the Habits of Certain other Anthophorinæ. Trans. Amer. Ent. Soc. 48, 1922, p. 137–156. 1 pl.; **Gehrs, C.**, Ueber den Nestbau von *Osmia spinulosa* K. Zeitschr. syst. Hymen. Dipt. 2, 1902, p. 4; **Graenicher, S.**, The Relations of the Andrenine Bees to the Entomophilous Flora of Milwaukee County. Trans. Wis. Acad. Sc. 15, 1905, p. 89–97; **Grossbeck, J. A.**, A Contribution toward the Life History of *Emphor bombiformis* Cress. Journ. N. Y. Ent. Soc. 19, 1911, p. 238–244, 1 pl. 2 figs.; **Hacker, L.**, Zur Biologie von *Megachile maritima*, einer Blattschneiderbiene. Natur. u. Offenb. 48, 1902, p. 94; **Hacker, H.**, Notes on the Genus Megachile and some rare insects collected during 1913–14. Mem. Queensl. Mus. 3, 1915, p. 137–141; **Heselhaus, F.**, Die Hautdrüsen der Apiden und verwandten Formen. Zool. Jahrb. Anat. 43, 1922, p. 369–464, 11 pls. **Höppner, H.**, Zur Biologie der Gattung Prosopis. Jahrb. Ver. Nat. Unterweser 1900, p. 56–58; Weitere Beiträge zur Biologie nordwestdeutscher Hymenopteren III. *Prosopis Kriechbaumeri* Förster. Allg. Zeitschr. Ent. 6, 1901, p. 291–293; **von Ihering, H.**, Biologia des abelhas solitarias do Brazil. Rev. Mus. Paulista 6, 1904, p. 461–481, 4 figs.; **Klein, E. J.**, Die Rosenblattbiene (*Megachile centuncularis*). Fauna Luxemburg 9, 1899, p. 76–86, 10 figs.; **Knab, F.**, How Emphor drinks. Proc. Ent. Soc. Wash. 13, 1911, p. 170; **Latreille, P. A.**, Observation sur l'abeille tapissière de Réaumur. Bull. Soc. Philom. 2, 1799, p. 33, fig.; Observations sur l'Abeille pariétine de Fabricius et considérations sur le genre auquel elle se rapporte. Ann. Mus. d'Hist. Nat. 3, 1804, p. 251–259, fig.; **Ludwig**, Nest und Vorratskammern der Loñalap von Ponape. Allg. Zeitschr. Ent. 9, 1904, p. 225–227, 1 fig. (*Megachile lonalap*); **Melander, A. L.**, The Nesting Habits of Anthidium. Biol. Bull. 3, 1902, p. 27–34, 10 figs.; **Melander, A. L.**, and **Brues, C. T.**,

Guests and Parasites of the Burrowing Bee Halictus. Biol. Bull. 5, 1903, p. 1–27, 7 figs.; **Müller, H.**, Ein Beitrag zur Lebensgeschichte der *Dasypoda hirtipes*. Verh. Nat. Ver. Preuss. Rheinl. Westph. 41, 1884, p. 1; **Müller, M.**, Zur Giologie unserer Apiden, insbes. der markischen Osmien. Zeitschr. wiss. Insektenbiol. 3, 1907–08, p. 247–251, 280–285; **Newberry, Miss M.**, Notes on the Nesting of *Anthidium paroselæ* Ckll. Psyche 9, 1900, p. 94; **Newport, G.**, On the Habits of *Megachile centuncularis*. Trans. Ent. Soc. London 4, 1845, p. 1–3; **Nichols, M. L.**, Some Observations on the Nesting Habits of the Mining Bee, *Emphor fuscojubatus* Ckll. Psyche 20, 1911, p. 107–112; **Nininger, H. H.**, Notes on the Life-history of *Anthophora stanfordiana*. Psyche 27, 1920, p. 135–137; **Pigeot, P.**, Note sur *Andrena nycthemera* Imh. Bull. Soc. Hist. Nat. Ardennes 10, 1903, p. 45–47; **Rau, P. and N.**, Notes on the Behavior of Certain Solitary Bees. Journ. Anim. Behav. 6, 1916, p. 367–370; **Sajó, K.**, Die Blattschneiderei der Megachile-Arten. Ill. Wochenschr. Ent. 1, 1896, p. 581–584, 2 figs.; **Schulz, W. A.**, Zur Kenntnis der Nistweise von *Euglossa cordata* (L). Allg. Zeitschr. Ent. 7, 1902, p. 153–154; **Schrottky, C.**, Die Nestanlage der Bienengattung Ptiloglossa Sm. Zeitschr. wiss. Insektenbiol. 2, 1906, p. 323–325; **Smith, J. B.**, Notes on Some Digger Bees. Journ. N. Y. Ent. Soc. 9, 1901, p. 29–40; II., p. 52–72, 3 pls., 3 figs.; **Spinola, M.**, Mémoire sur les Mœurs de la Cératine albilabre. Ann. Mus. d'Hist. Nat. 9, 1807, p. 236–248; **Stöckert, F. K.**, Zur Biologie von *Prosopis variegata* F. Konowia 1, 1922, p. 39–58; **Strand, E.**, Beschreibung je einer neuen Allodape- und Ceratina-Art aus Kammerun, nebst biologischen Bemerkungen. Ent. Mitt. 3, 1914, p. 173–176; **Strohl, J.**, Die Copulationsanhänge der solitären Apiden und die Artstehung durch "physiologische Itolierung." Inaug. Diss. Freiburg i. Br. 1908, 52 pp., 3 pls., 2 text-figs. **Turner, C. H.**, The Homing of the Burrowing Bees (Anthophoridæ). Biol. Bull. 15, 1908, p. 247–258; The Sun-dance of Melissodes. Psyche 15, 1908, p. 122–124; **Vachal, J.**, Les insectes actuels temoins des revolutions du globe. Bull. Soc. Ent. France 1905, p. 68–70 (Andrena); Sur les abeilles (Apidæ) de la période glaciaire. *ibid.* 1906, p. 131–134 (*Andrena gentianæ*); **Verhoeff, C.**, Zur Lebensgeschichte der Gattung Halictus (Anthophila), insbesondere einer Uebergangsform zu socialen Bienen. Zool. Anzeig. 20, 1897, p. 369–393, 21 figs.; **Walsh, B. D.**, Mason Bees and their Habits. Amer. Ent. 1, 1868, p. 8–11; **Wesenberg-Lund, C.**, Traek af Linne's Vaegge Bi's (*Anthophora parietina*). Ent. Medd. 2, 1899 p. 97–120, 1 pl.; **Xambeu, V.**, Mœurs et métamorphoses des *Anthidium oblongatum* et 7–*dentatum*, Hyménoptères du groupe des Apides. Bull. Soc. Ent. France 1896, p. 328–331; **Zollinger, H.**,

324   SOCIAL LIFE AMONG THE INSECTS

Beiträge zur Naturgeschichte der Holzhummel (*Xylocopa violacea amethystina*). Nat. en Geneesk. Arch. v. Neerlands Indie 3, 1846, p. 295. Transl. by Franzius, Stett. Ent. Zeitg. 12, 1851, p. 236–240. SOLITARY BEES (Parasitic). **Alfken, J. D.**, Die Nomada-Arten Nordwest-Deutschlands als Schmarotzer. Zeitscher. Syst. Hymen. Dipt. 2, 1902, p. 5–10; **Armbruster, L.**, Zur Phylogenie, etc., *l.c.*, 1916 (Sphecodes); **Breitenbach, W.**, Ueber *Halictus 4-cinctus* F. und *Sphecodes gibbus* L. Stettin. Ent. Zeitg. 1878, p. 241–243; **Bréthes, J.**, Una Anthophorina parasita? Anal. Mus. Nac. Buenos Aires (3) 12, 1909, p. 81–83; **Davidson, A.**, *Alcidamia producta* Cress. and its Parasites. Ent. News 7, 1896, p. 216–218, 1 fig. (*Stelis 6-maculata*); **Ferton, C.**, Sur les Mœurs des Sphecodes Latr. et des Halictus Latr. Bull. Soc. Ent. France 1898, p. 75–77; Notes détach. *l.c.* 2, p. 503–504, and 8, p. 89–93 (Dioxys); 3, p. 60–61 (Sphecodes); 4, p. 551 (Cœlioxys), 9, p. 332–333 (Crocisa); *Perezia maura*, nouveau genre d'Apiaires parasites d'Algérie. Ann. Soc. Ent. France 83, 1914, p. 233–237, 1 fig.; **Fox, W. J.**, *Cœlioxys 8-dentata* and its host Ent. News 11, 1900, p. 553; **Graenicher, S.**, Some Obesrvations on the Life History and Habits of Parasitic Bees. Bull. Wis. Nat. Hist. Soc. 3, 1905, p. 153–167, 1 pl.; **Hoffer, E.**, Ueber die Kuckucksbienen. Mitt. Nat. Ver. Steiermark 43, 1907, p. 431–433; **Johnson, J. C.**, A Case of Parasitism of *Melecta armata* on *Anthophora acervorum*. Zoologist (4) 17, 1913, p. 427–429; **Latter, O. H.**, How do Inquiline Bees find the Nest of their Host? Nature 74, 1906, p. 200; **Marchal, P.**, Le Parasitisme des Sphecodes. Bull. Soc. Ent. France 63, 1894, p. CXV.; **Morice, F. D.**, The probable " hosts " of *Ammobates carinatus* Morawitz and *Phiarus melectoides* Smith. Zeitschr. syst. Hymen. Dipt. 3, 1903, p. 317; **Newport, G.**, Note on the parasitic Habits of Nomadæ. Trans. Ent. Soc. London 4, 1842 Proc., p. 67; **Nielsen, J. C.**, Om Bislaegeten Sphecodes, Latr. Ent. Medd. (2) 2, p. 22–28, English résumé, p. 29–30; **Pérez, J.**, Les Apiares Parasites au point de Vue de la Théorie de l'Évolution. Bordeaux 1884; Sur les Mellifères parasites. Act. Soc. Linn. Bordeaux 50, 1897, p. XV–XVII; **Perkins, R. C. L.**, Notes on Some Habits of Sphecodes Latr. and Nomada Fabr. Ent. Month. Mag. 23, 1887, p. 271–274; Is Sphecodes parasitic? *ibid.* 25, 1889, p. 206–208; **Rabaud, E.**, Note pour servir a l'étude psychologique du mimétisme. Feuille jeun. Natural (5) 41, 1911, p. 159; **Semichon, L.**, Sur la ponte de la *Melecta armata* Panzer. Bull. Soc. Ent. France 1904, p. 188–189; Les Conditions d'existence de *Melecta armata* Panz. a l'état d'œuf et de larve. *ibid.* 1911, p. 372–374; Parasitisme provoqué entre deux larves d'*Halictus quadricinctus* Fabricius. *ibid.* 1912, p. 90–92; **Sladen, F. W. L.**, Observations on *Sphe-*

codes rubicundus v. Hag. Ent. Month. Mag. 31, 1895, p. 256;
Smith, F., Notes on the Parasites of the Genus Nomada and on
Other Insects. Trans. Ent. Soc. London 3, 1843, p. 293–294;
Spinola, M., Note sur les Stélides. Rev. Zool. 2, 1839, p. 305–307,
334; **Turner, C. H.**, Notes on the Behavior of a Parasitic Bee of
the Family Stelidæ. Journ. Anim. Behav. 1, 1911, p. 374–392;
Verhoeff, C., Zur Kenntnis des biologischen Verhältnisses zwischen
Wirt- und Parasiten-Bienenlarven. Zool. Anzeig. 15, 1892, p. 41–43;
Wheeler, W. M., The Parasitic Aculeata, a Study in Evolution. Proc.
Amer. Phil. Soc. 58, 1919, p. 1–41.

5. When this paragraph on the Halicti was written I had not seen
Armbruster's paper (Zur Phylogenie der Geschlechtsbestimmungs-
weise bei Bienen *l.c.* 1916), which contains new and interesting observations on the alternation of generations, based on Alfken's and
von Wagner's flight-records of some 35 species in Germany. The
records show that all the North European forms have two generations
like those observed by Fabre in Southern France (1879–80, Souvenirs
8, Chapt. 7–9), but that the female of the sexual generation, which
appears in July and August rarely or never hibernates, but dies after
provisioning its brood. The larvæ, which passes the winter, give rise
in late March or early April to the spring brood consisting entirely of
females, and these disappear in June or July after producing from their
unfertilized eggs the males and females of the fall generation. Alfken's
and von Wagner's records of some 17 species of Sphecodes, which are
parasitic on the Halicti, show the very same phenomena. From
observations which I have been making for several years, I am convinced that our small metallic green New England Halicti (subgenus
Chloralictus) exhibit essentially the same conditions as the German
species. It would be interesting to observe the behavior of the autumn
generation of these and other Halicti in our Southern States, where
the climatic conditions resemble those of Southern France and where
we might expect, therefore, the adult females to hibernate and start
the parthenogenetic generation the following spring. Alfken and Armbruster call attention to the fact that this generation consists of a much
greater number of individuals than the sexual generation. This is
also true of the females of Chloralictus, which may be very scarce,
though the males of the same generation are often abundant during
August. Since both arise from unfertilized eggs, it is evident that
these, in Chloralictus at least, have a very pronounced tendency to
produce males.

6. The following works comprise most of the important contributions
to our knowledge of the bumble-bees (Bombus) and their parasites

(Psithyrus); **Armbruster, L.**, Probleme des Hummelstaates. Biol. Centralbl. 34, 1914, p. 685–707, 2 pls.; **Bachmann, M.**, Beobachtungen vor dem Hummelnest. Ent. Zeitschr. Frankf. a.M. 29, 1916, p. 89–90, 93–94, 98–99, 103–104; 30, 1–3; **Benetti, V.**, Richerche biologiche sui Bombi. Monit. Zool. Ital. 13, Suppl. 1902, p. 38–40; **Bengtsson, S.**, Studier och iakttagelser öfver Humlor. Ark. Zool. 1, 1903, p. 197–222, **Bethune, C. J. S.**, Some Observations on a Bumble-bees' Nest. 30th Ann. Rep. Ent. Soc. Ontario (1899) 1900, p. 111–112; **Burnett, W. I.**, Facts observed in the generation of the Humble Bee (*Bombus americanus*) and Aphides, as illustrating some obscure phenomena in the physiology of generation. Proc. Boston Soc. Nat. Hist. 3, 1850, p. 262–264; **von Buttel-Reepen, H.**, Die stammesgeschichtliche Entstehung des Bienenstaates, etc., *l.c.* 1903; Die Lebensweise der Hummeln. Nat. Wochenschr. 19, 1904, p. 299–300; Zur Psychobiologie der Hummeln I. Biol. Centralbl. 27, 1907, p. 579–587, 604–613; **Coville, F. V.**, Notes on Bumble-Bees. Proc. Ent. Soc. Wash. 1, 1890, p. 197–203; **Dreyling, L.**, Beobachtungen über die wachsabscheidenden Organe bei den Hummeln. nebst Bemerkungen über die homologen Organe bei Trigonen. Zool. Anzeig. 29, 1905, p. 563–573, 6 figs.; **Dupetit-Thouars, L. M. A.**, Sur quelques habitudes des abeilles-bourdons. Nouv. Bull. Soc. Philom. 1, 1807, p. 45; **Erikson, E. V.**, On the Psychology of Bumble-bees (Russian). Rev. Russe Ent. 8, 1908, p. 32–41; **Fairchild, D.**, and **Barrett, O. W.**, Notes on the Copulation of *Bombus fervidus*. Proc. Ent. Soc. Wash. 8, 1906, p. 13–14, 1 pl.; **Ferton, C.**, Notes détach. *l.c.* 1, p. 84–85, 4, p. 536–537 (Bombus); **Förster, F.**, Zur Schwirrbewegung der Bienen im Stocke. Ent. Wochenbl. 25, 1908, p. 85; **Franklin, H. J.**, The Bombidæ of the New World. Trans. Amer. Ent. Soc. 38, 1912, p. 177–486; 39, 1913, p. 73–200, 22 pls.; **Friese, H.**, Die Schmarotzerbienen und ihre Wirte. Zool. Jahrb. Syst. 3, 1888, p. 847–870 (Psithyrus); Ueber Hummelleben im arktischen Gebiete. Allg. Zeitschr. Ent. 9, 1904, p. 409–414, 1 fig.; **Friese, H.** and **von Wagner, F.**, Zoologische Studien an Hummeln. Zool. Jahrb. Syst. 29, 1910, p. 1–104, 7 pls.; **Frionnet, C.**, Bombus et Psithyrus de France et de Belgique. Feuille jeun. Natural. (4) 32, 1902, p. 165–169, 177–183, 1 pl.; **Frison, T. H.**, Note on the Habits of *Psithyrus variabilis* Cress. Bull. Brooklyn Ent. Soc. 11, 1916, p. 46–47; Notes on Bombidæ, and on the Life History of *Bombus auricomus* Robt. Ann. Ent. Soc. Amer. 10, 1917, p. 277–286, 2 pls.; Additional Notes on the Life History of *Bombus auricomus*. *ibid.* 11, 1918, p. 43–48, 3 pls.; Keys for the Separation of the Bremidæ, or Bumblebees of Illinois and Other Notes. Trans. Ill. State Acad. Sc. 12, 1919, p. 157–166; *Psithyrus laboriosus* Fabr. in the Nests of Bumblebees. Canad. Ent. 53, 1921, p. 100–

101; Report on the Bremidæ collected by the Crocker Land Expedition. Bull. Amer. Mus. Nat. Hist. 41, 1919, p. 451–459, 1 pl., 2 textfigs.; **Giard, A.**, Sur une particularité éthologique de *Bombus confusus* Schenck. Bull. Soc. Ent. France 1899, p. 82–83; **Graeffe, E.**, Beobachtungen an Hummelarten der Alpen Steiermarks. Mitt. nat. Ver. Steiermark 48, 1912, p. 376–380; **Gundermann, E.**, Einige Beobachtungen an Hummelnestern. Ent. Wochenbl. 25, 1908, p. 30–31, 35–36; **Härter, R.**, Biologische Beobachtungen an Hummeln, 27. Ber. Oberhess. Ges. Natur. u. Heilk. 1890, p. 59–75; **Havenhorst, P.**, De paring van *Bombus hortorum* L. Ent. Bericht. 2, 1909, p. 305; **Heikertinger, F.**, Der Rotklee und die Hummeln. Zeitschr. naturw. erdkundl. Unterr. 1919, p. 374–376; **Höppner, H.**, Ueber zwei unbekannte oder wenig bekannte Hummelnester. Ent. Nachr. 23, 1897, p. 313–316; Weitere Beiträge zur Biologie nordwestdeutscher Hymenopteren II. Ueber das Vorkommen mehrerer Bombus-Arten in einem Neste. Allg. Zeitschr. Ent. 6, 1901, p. 132–134, 1 fig.; **Hoffer, E.**, Ueber die Lebensweise des *Apathus* (*Psithyrus*) *campestris*. Mitt. naturw. Ver. Steiermark 1881, p. 87–92; Die Hummeln Steiermarks. Lebensgeschichte und Beschreibung derselben. Graz, Leuschner u. Lubensky 1882–1883; Neue Hummelnester von den Hochalpen. Kosmos 1, 1885, p. 291–300; Die Schmarotzerhummeln Steiermarks. Lebensgeschichte und Beschreibung derselben. Mitt. Naturw. Ver. Steiermark 25, 1888, p. 82–159, 1 pl.; Ueber den sogenannten Trompeter in den Hummelnestern. *ibid.* 1905, p. LVIII–LIX; **von Ihering, R.**, Biologische Beobachtungen an brasilianischen Bombus-Nestern. Allg. Zeitschr. Ent. 8, 1903, p. 447–453, 5 figs.; Zur Frage nach dem Ursprung der Staatenbildung bei den sozialen Hymenopteren. Zool. Anzeig. 27, 1903, p. 113–118; **Jarvis, T. D.**, Bumblebees that Fertilize the Red Clover. 36th Ann. Rep. Ent. Soc. Ontario 1906, p. 128–129; **Kristof, L.**, Einige Beobachtungen über das Leben einheimischer Hummeln, etc. Mitt. Naturw. Ver. Steiermark. 1883, p. LXIV–LXXIV; **Kultscher, A.**, Der Trompeter im Hummel-Staate. Ill. Wochenschr. Ent. 1, 1896, p. 271–274; **Labillardière, J. J. H.**, Note sur les mœurs des Bourdons. Mém. Mus. d'Hist. Nat. 1, 1815, p. 55–59; **Lie-Pettersen, O. J.**, Bidrag til kindskaben om Vestlandets Bombus-og Psithyrus-arter. Bergens Mus. Aar. 1900, 19 pp.; Biologische Beobachtungen an norwegischen Hummeln. *ibid.* 1901, 10 pp.; Neue Beiträge zur Biologie der norwegischen Hummeln. *ibid.* 1906, 41 pp.; 1 fig.; **Lindhard, E.**, Om Roedkloeverens Bestoevning og de Humlebiarter, som herved er virksomme. Tidskr. Landbrug. Planteavl 18, 1911, p. 719–737, 2 figs.; Humlebien som Husdyr. *ibid.* 19, 1912, p. 335–351, 4 figs.; Om Roedkloeverracer med kort Kronroer og blomsterbesoegende Bier, *ibid.* 27, 1920, p. 653–680, 5 figs.; **Lutz, F. E.**, The Geographic Distribution of Bombidæ (Hymenoptera) with Notes on Certain Species of Boreal America. Bull. Amer. Mus. Nat. Hist. 35, 1916, p. 500–521, 1 fig.; **Murchardt, H.**, Bidrag till humlornas och synlthumlornas utbredning. Ent. Tidskr. 25, 1904, p. 204; **Pérez, J.**, (Curieuse habitude du *Bombus lefebvrei*). Act. Soc. Linn. Bordeaux 52, 1897, p. LXXXVII–LXXXVIII; **Plath, O. E.**, Notes on Psithyrus, with Records of two new American Hosts. Biol. Bull. 43, 1922, p. 23–44, 1 pl.; A Unique Method of Defense of *Bremus* (*Bombus*) *fervidus* Fabricius. Psyche 29, 1922, p. 180–185; **Plath, O. E.**, Notes on the Nesting Habits of Several North American Bumblebees. Psyche 29, 1922, p. 189–202. **Putnam, F. W.**, Notes on the Habits of Some Species of Humble-Bees. Proc. Essex Inst. Salem, Mass., 4, 1865, p. 98–105; **Purovsky, R.**, Auffallendes Vorkommen eines Hummelnestes. Ill. Zeitschr. Ent. 4, 1899, p. 123; **de Réaumur, R. A.**, Mémoires pour servir a l'Histoire des Insectes. Paris, Imp. Royale 6, 1742, pp. 1–38, 4 pls.; **Reichert, A.**, Auffallendes Vorkommen von Hummelnestern I. Ill. Zeitschr. Ent. 4, 1899, p. 283; **Rothe, H. H.**, Das Leben der Hummeln. Naturw. Wochenschr. 18, 1903, p. 457–462; **Saunders, E.**, Notes on a Nest of *Bombus hortorum* race *subterraneus*. Ent. Month. Mag. (2) 9, 1898, p. 250–251; **Schulz, W. A.**, Ueber das Nest von *Bombus cayennensis* (L.). Verh. zool. bot. Ges. Wien 51, 1901, p. 361–362; Nachtrag, *ibid.*, p. 762; **Sladen, F. W. L.**, Bombi in Captivity, and Habits of Psithyrus. Ent. Month. Mag. (2) 10, 1899, p. 230–234; The Humble-Bee, its Life-History and How to Domesticate It, with descriptions of all the British Species of Bombus and Psithyrus. London, Macmillan, 1912; Inquiline Bumble-bees in British Columbia. Canad. Ent. 47, 1915, p. 84; The Wasps and Bees Collected by the Canadian Arctic Expedition, 1913–18. Rep. Canad. Arct. Exped. 3. Insects 1919, p. 25G–35G, 2 pls.; **Smith, F.**, Notes on the Nest of *Bombus Deshamellus*. Trans. Ent. Soc. London (2) 1, 1851, Proc. p. 111–112; **Sparre-Schneider, J.**, Humlerne og deres forhold till flora en i det arktiska Norge. Tromsö Mus. Aarhefter 1894, p. 133; Hymenoptera aculeata im arktischen Norwegen. Tromsö 1909; **Stierlin, R.**, Ueber das Leben der Hummeln. Mitt. Nat. Ges. Winterthur 1906, p. 130–144; **Sundvik, E. E.**, Ueber das Wachs der Hummeln. Zeitschr. physiol. Chem. 26, 1898, p. 56–58; Biologiska iakttagelser angäende humlorna (humleväger). Medd. Soc. F. et Fl. Fenn. 32, 1906, p. 156; Ueber das Wachs der Hummeln II. Mitteilung. Psyllaalkohol, ein Bestandteil des Hummelwachses. Zeitschr. phys. Chem. 53, 1907, p. 365–369; Biologiska iakttagelser i afseende a humlorna. Medd. Soc. F. et Fl. Fenn. 34, 1908, p. 131; Iakttagelser angäende humlorna. *ibid.* 37, 1911, p. 56; **Tuck, W. H.**, Note on the habits of *Bombus Latreillellus.* Ent. Month. Mag. (2) 8, 1897, p. 234–235; **Wagner, W.**, Psychobiologische Untersuchungen an Hummeln. Erster Teil. Zoologica 19, 1906, 78 pp., 1 pl., 50 figs.; **Zweiter** Tiel. *ibid.* 19, 1907, p. 79–239, 86 figs.; **Westerlund, A.**, Wie Bombus seinen Nestbau beginnt. Ill. Zeitschr. Ent. 3, 1898, p. 113–114.

7. With the exception of a few very early papers, the following bibliography comprises nearly all that has been written on the habits of the Meliponinæ: **Bennett, E. T.**, Mexican Bees (*Melipona Beechii*) in F. W. Beechey, Narrative of a Voyage to the Pacific, etc. London 1831, Vol. 2, p. 613–618, fig.; **Bigg, H.**, On a Species of Bee from the Brazils, found living on splitting a log of peach wood containing the comb. Proc. Zool. Soc. London 2, 1834, p. 118; **von Buttel-Reepen, H.**, Die stammesgeschichtliche Entwicklung, etc. *l.c.* 1903; *Trigona emerina* F. in Psychobiologische und biologische Beobachtungen *l.c.* 1907; **du Buysson, R.**, Sur deux Mélipones (Hyménoptères) du Mexique. Bull. Mus. d'Hist. Nat. 1901, p. 104–106; Sur deux Mélipones du Mexique. Ann. Soc. Ent. France 70, 1901, p. 153–156, 2 pls.; **Coelho de Scabra, V.**, Noticia de diversas Abelhas que dao mel proprias do Brasil o desconhecidas na Europa. Mem. Acad. Sc. Lisboa 2, 1799, p. 90–104; **Dreyling, L.**, Zur Kenntnis der Wachsabscheidung bei Meliponen. Zool. Anzeig. 28, 1904, p. 204–210, 2 figs.; Beobachtungen über die wachsabscheidenden Organe bei den Hummeln, etc., *l.c.* 1905; **Drory, E.**, Quelques, Observations sur la Mélipone scutellaire. Bordeaux 1872; Einige Beobachtungen an *Melipona scutellaris.* Bienenzeitung, Nördlingen 28, 1872, p. 13–18; Ueber Meliponen. Bienenzeitg. 1873, p. 172–176; Notes sur quelques espèces de Mélipones de l'Amérique du Sud. C. R. Soc. Linn. Bordeaux 29, 1873, p. 29; Nouvelles Observations sur les Mélipones, Le Rucher du Sud-Ouest, Bordeaux 1, 1873; De la Manière dont les Mélipones secrètent la cire. C. R. Soc. Linn. Bordeaux 29, 1873, p. 62; Welchen wissenchaftlichen und praktischen Werth haben die Meliponen in Europa? Bienenzeitg. 30, 1874; **Ducke, A.**, Die stachellosen Bienen (Melipona Ill.) von Pará, nach dem Materiale der Sammlung des Museu Goeldi beschrieben. Zool. Jahrb. Syst. 17, 1902, p. 285–328, 1 pl.; **Fiebrig, K.**, Skizzen aus dem Leben einer Melipone aus Paraguay. Zeitschr. wiss. Insektenbiol. 3, 1908, p. 374–386, 3 figs.; **Giard, M.**, Notes sur les Mœurs des Mélipones et des Trigones du Brésil. Ann. Soc. Ent. France (5) 4, 1874, p. 567–573; **Hamlyn-Harris, R.**, The Stingless Bees of North and South America considered in the Light of Domestication. Ent. Rec. Journ. Var. 15, 1903, p. 99–100; **von Ihering, H.**, Der Stachel der Meliponen. Ent. Nachr.

12, 1886; Observation biologique. Bull. Soc. Ent. France 1902, p. 23 (*Trigona helleri* living in termitaria); Biologie der Stachellosen Honigbienen Brasiliens. Zool. Jahrb. Syst. 19, 1904, p. 79–287, 13 pls., 8 figs.; A producçao de cera e mel em nossas abelhas indigenas (Meliponidæ). Ent. Brasiliero 2, 1909, p. 164–167, 1 fig.; Phylogenie der Honigbienen. Zool. Anzeig. 38, 1911, p. 129–136, 1 fig.; Zur Biologie der brasilianischen Meliponiden. Zeitschr. wiss. Insektenbiol. 8, 1912, p. 1–5, 43–46; Karawaiew, W., The Nestorifice of *Trigona apicalis* (Russian). Kiew 1902, p. 10–11; Lutz, F. E., and Cockerell, T. D. A., Notes on the Distribution and Bibliography of North American Bees of the Families Apidæ, Meliponidæ, Bombidæ, Euglossidæ and Anthophoridæ. Bull. Amer. Mus. Nat. Hist. 42, 1920, p. 491–641; Marshall, W., Die stachellosen Bienen Südamerikas. Leipz. Bienenzeitg. Heft. 9, 1898, Müller, F., Poey's Beobachtungen über die Naturgeschichte der Honigbiene in Cuba, *Melipona fulvipes* Guér. Im Auszug und mit Anmerkungen. Zool. Garten, Frankf. a.M. 16, 1875, p. 291–297; Stachellose brasilianische Honigbienen zur Einführung in zoologische Gärten empholen. *ibid.* 16, 1875, p. 41–55; Nurse, C. G., A New Species of Indian Wax-producing Bee. Journ. Bombay. Nat. Hist. Soc. 17, 1907, p. 619 (*Melipona cacciæ*); Parish, C. S. P., Pwai-ngyet (*Trigona lœviceps*) Science Gossip 1866, p. 198–200, 2 figs.; Peckholt, T., Ueber brasilianische Bienen. Die Natur. 42, 1893, p. 579–581; 43, 1894, p. 87–91, 223–225, 233–234; Pérez, J., On the Production of Males and Females in Melipona and Trigona. Ann. Mag. Nat. Hist. (6) 16, 1895, p. 125–127 (Transl. from C. R. Acad. Sc. Paris 120, 1895, p. 273–275); Prell, Menschenschädel als Bienenwohnung. Ent. Mitt. 8, 1919, p. 157–162, 6 figs. (*Trigona canifrons*); Rudow, Die Wohungen der honigsammelnden Bienen (Anthophilidæ). Ent. Zeitschr. 27, p. 1–19, 19 figs.; Smith, F., On *Trigona lœviceps*. Trans. Ent. Soc. London (2) 4, 1858, p. 98–99; Schulz, W. A., Neue Beobachtungen an südbrasilianischen Meliponiden-Nestern. Zeitschr. wiss. Insektenbiol. 1, 1905, p. 199–204, 250–254, 6 figs.; Die indoaustralische *Trigona lœviceps* Sm. und ihr Nest. *ibid.* 3, 1907, p. 65–73, 4 figs.; Ein javanisches Nest von *Trigona canifrons* F. Sm. in einem Bambusstabe. *ibid.* 5, 1909, p. 338–341, 5 figs.; Silvestri, F., Contribuzione alla Conoscenza dei Meliponidi del Bacino del Rio de la Plata. Riv. Patol. Veget. 10, 1902, p. 121–170, 3 pls., 19 figs.; Spinola, M., Observations sur les apiaires Méliponides. Ann. Sc. Nat. (2) 13, 1840, p. 116–140, fig.; Note sur les Hyménoptères de la tribu de Méliponides. Rev. Zool. 5, 1842, p. 216–218, 267–268; Strand, E., Biologische Notiz über papuanische Trigonen. Intern. Ent. Zeitschr. Guben 6, 1912, p. 11; Tomaschek, Ein Schwarm der amerikanischen Bienenart *Trigona lineata* (?) lebend in Europa. Zool. Anzeig. 2, 1879, p. 582–587; 3, 1880, p. 60–65; **Wasmann, E.**, Ein neurer Melipona-Gast (*Scotocryptus Goeldii*) aus Pará. Deutsch. Ent. Zeitschr. 1899, p. 411; Contribuiçao para o estudo dos hospedes do abelhas brazilieras. Rev. Mus. Paulista 6, 1904, p. 482–487, 1 pl.; **Waterhouse, C. O.**, Notes on the Nests of Bees of the genus Trigona. Trans. Ent. Soc. London 1903, p. 133–136, 1 pl., 3 figs.; **Wheeler, W. M.**, Notes on the Habits of Some Central American Stingless Bees. Psyche 20, 1913, p. 1–9.

8. For an account of these fossil bees see **von Buttel-Reepen,** Leben und Wesen der Bienen. Braunschweig, Viehweg. u. Sohn 1915, p. 7–14 and his "Apistica" (*vide infra*).

9. Of the following literature dealing with the various species of Apis and the varieties of the honey bee, von Buttel-Reepen's "Apistica" is the most important. It contains a reprint of Gerstäcker's rare and valuable paper of 1862 (p. 124–154): **Benton, F.**, (Proposed Introduction of *Apis dorsata* into the United States). Proc. Ent. Soc. Washington 4, 1896, p. 36; The Importation and Breeding of Honey Bees of Various Types. Bur. Ent. U. S. Dep. Agric. 52, 1905, p. 103–108; **Butlerow, A.**, Die kaukasische Biene und die Bienenzucht im Kaukasus. Bienenzeitg. Eichstadt, 1878, p. 34; Mitteilungen über die kaukasische Biene und deren Zucht in Deutschland. *ibid.* 1879, p. 311; **von Buttel-Peepen, H.**, Die stammesgeschichtliche Entstehung, etc., *l. c.* 1903; Apistica. Beiträge zur Systematik, Biologie, sowie zur geschichtlichen und geographischen Verbreitung der Honigbiene (*Apis mellifica* L.). Mitt. zool. Mus. Berlin 3, 1906, p. 117–201, 8 figs.; Zur Biologie von *Apis indica* F. in Psychobiologische und biologische Beobachtungen etc. *l.c.* 1907 (Jacobson's observations in Java); Entomologischer Reisebrief aus Ceylon's Bergen. Ent. Mitt. 1, 1912, p. 97–103, 1 fig. (*Apis indica*). Leben und Wesen der Bienen. *l.c.* 1915; **Dathe, R.**, Apis dorsata. Bienenzeitg. Nördlingen 1883, No. 19, p. 234; **Enderlein, G.**, Neue Honigbienen und Beiträge zur Kenntnis der Verbreitung der Gattung Apis. Stett. Ent. Zeitg. 67, 1906, p. 331–344, 4 figs.; **Friese, H.**, Ueber den Waben-Bau der indischen Apis-Arten. Allg. Zeitschr. Ent. 7, 1902, p. 198–200, 2 figs.; **Gerstäcker, A.**, Ueber die geographische Verbreitung und die Abänderungen der Honigbienen nebst Bemerkungen über die ausländischen Honigbienen der alten Welt. Festschr. 11. Wandervers. Deutsch. Bienenwirte zu Potsdam 1862 (Reprinted in von Buttel-Reepen's "Apistica") Zur geographischen Verbreitung der Honigbiene. Nachträg. Stettin. Ent. Zeitg. 1864; Ueber die geographische Verbreitung der Honigbiene. Bienenzeitg. Eichstädt. 1866, p. 60–65;

Gravenhorst, C. J. H., Die cyprische Biene. *ibid.* 1877, p. 73; **Hamlyn-Harris, R.,** *Apis dorsata* Fabr. considered in the light of Domestication. Ent. Rec. Journ. Var. 14, 1902, p. 12–14; **Helms, R.,** Races and Relationships of Honey Bees. Zoologist (3) 20, 1896, p. 201–209 (from Agric. Gaz. N. S. Wales); **Latreille, P. A.,** Notice des espèces d'Abeilles vivant en grande société, ou Abeilles proprement dites, et descriptions d'espèces nouvelles. Ann. Mus. d'Hist. Nat. 5, 1804, p. 161–178; Mémoire sur un gateau de ruche d'une Abeille des Grandes-Indes, et sur les différences des abeilles proprement dites, vivant en grande société, de l'ancien continent et du nouveau. *ibid.* 4, 1804, p. 383–394, fig.; **Musy, M.,** *L'Apis dorsata* de Sumatra. Bull. Soc. Fribourg Sc. Nat. 17, 1910, p. 103–104; **Newman, E.,** South African Honey-Bee. Zoologist 1885, p. 4675; **Noll,** Notiz über *Apis fasciata.* Bienenzeitg. Eichstädt 1867, p. 128; **Rothschütz, E.,** Miscellanea über *Apis indica, Melipona minuta* und die krainer Biene. *ibid.* 1872, p. 40; **Schneider, G.,** Ueber eine Urwald-Biene (*Apis dorsata* F.). Zeitschr. wiss. Insektenbiol. 4, 1908, p. 447–453, 2 figs.; **von Siebold, C.,** Ein Wort über die ägyptischen wahren Drohnenmütter. Bienenzeitg. Nördlingen 1865, No. 1, p. 8–9; **Vogel, F. W.,** Die ägyptische Biene (*Apis fasciata*), ihre Einführung durch den Akklimatationsverein zu Berlin. Berlin 1865, 58 pp.; Die ägyptische Biene III, Bienenzeitg. Nördligen 1865, No. 1, p. 5–8; **Vosseler, J.,** Die ostafrikanische Honigbiene. Ber. Land-Forstwirtsch. Deutsch-Ostafr. 2, 1905, p. 15–29.

10. Owing to the vastness of the literature on the behavior of the honey-bee it is quite impossible within the scope of the present volume to present even a selected bibliography that might be useful to the general reader. I can only refer for adequate citation of the older literature (up to 1862) to Hagen's Bibliotheca Entomologica and for that of more recent contributions of value to such works as von Buttel-Reepen's Leben und Wesen der Bienen and E. F. Phillips' Beekeeping, New York, Macmillan 1915.

11. For accounts of the nesting of honey-bees in the open air, in tree-trunks and the soil, see the following papers: **Bouvier, E. L.,** Sur une nidification rémarquable d'*Apis mellifica* L. observée au Museum de Paris. Bull. Soc. Ent. France 1904, p. 187–188; (Suite et fin.) *ibid.* 1905, p. 144–145; Une colonie d'Abeilles. Bull. Soc. Nat. d'Agric. France 1904, p. 503, 504; Sur la nidification d'une colonie d'abeilles à l'air libre. Bul. Soc. Philom. Paris (9) 7, 1905, p. 186–206, 5 figs.; Observation sur le nid aërien figuré par Curtis. Bull. Soc. Ent. France 1905, p. 222, 223; La nidification des abeilles à l'air libre. C. R. Acad. Sc. Paris 142, 1906, p. 1015–1020; La nidification des abeilles à l'air libre. Le Cosmos N. S. 54, 1906, p. 636–638; Nouvelles Observa-

tions sur la nidification des Abeilles à l'air libre. Ann. Soc. Ent. France 75, 1906, p. 429–444, 3 pls.; Sur les nids aëriens de l'Abeille mellifique (nouveaux faits). Bull. Soc. Ent. France 1907, p. 294–296; **Clement, A. L.**, Une colonie d'abeilles au Jardin des Plantes. La Nature, 29, 1901, p. 204–206, 2 figs.; Une colonie d'abeilles au Jardin des Plantes. *ibid.* 29, p. 301–302, 1 fig.; Un essaim d'Abeilles au Museum. *ibid.* 1904, p. 218, 219, 1 fig.; **Curtis, J.**, British Entomology, Hymenoptera: *Apis mellifica,* the common Hive or Honey Bee. 1862, p. 769; **Ferton, C.**, Notes détach. *l.c.* 4, p. 535–536 (*A. mellifica* nesting in the soil); **Nibelle, M.**, Note sur un cas de nidification d'Abeille commune à l'air libre. Bull. Soc. Amis Sc. Nat. Rouen (5) 43, 1908, p. 25–27; **Gosse, P. H.**, Description of a Bee-tree. Zoologist 2, 1844, p. 607–609; **Höppner, H.**, Ein Freibau unserer Honigbiene (*Apis mellifica* L., Mitt. Nat. Mus. Crefeld 1913, p. 22–26, 1 pl.; **Schmidt, A.**, Zoologische und botanische Mitteilungen. Schr. Nat. Ges. Danzig N. F., 9, p. 94–96 (Nest in tree trunks); **Nielsen, J. C.**, Om fritbyggede Honningbireder i Danmark (On the Nidification of honey-bees in the open air in Denmark). Vidensk. Medd. Naturh. Foren. 64, 1912, p. 34–37, 4 pls.

12. *Cf.* the literature cited for the second lecture. The tendency of solitary wasps and bees to lay female-producing eggs first in the series was designated by Verhoeff as " proterothesis." It has been most recently observed by Taylor in Ancistrocerus (Notes on the Biology of Certain wasps of the Genus Ancistrocerus, etc., *l.c.* 1922). It should be noted also that the recently fecundated queen ant always conforms to this rule, since her first offspring are invariably workers and therefore females.

13. It would serve no useful purpose to cite the voluminous literature that has been recently produced in Germany by the cohorts of Ferd. Dickel and those of von Buttel-Reepen in their acrimonious controversy concerning the parthenogenetic or nonparthenogenetic origin of the male honey-bee. An impartial digest of the whole matter has not yet been made.

14. **Phillips, E. F.**, Beekeeping, *l.c.*, p. 187, 188.

15. *Cf.* **von Buttel-Reepen**, Ueber den gegenwärtigen Stand der Kenntnisse von den geschlechtsbestimmenden Ursachen bei der Honigbiene (*Apis mellifica* L.). Verh. Deutsch. Zool. Gesell. 1904, p. 65–66, *nota,* and Leben und Wesen der Bienen, *l.c.*, p. 42.

16. **Onions, G. W.**, South African " Fertile Worker Bees." Agric. Journ. Union S. Afr. 3, 1912, p. 720–728; *ibid.* 7, 1914, p. 44–46; **van Warmelo, D. G.**, South African Fertile Worker Bees and Parthenogenesis. *ibid.* 3, 1912, p. 786–789. Phillips, from whom these refer-

ences are taken, says that Onions's "claim is supported by considerable evidence," but that van Warmelo denies his statement. More recently **R. W. Jack** (Parthenogenesis amongst the Workers of the Cape Honey-Bee. Trans. Ent. Soc. London, 1917, pp. 396–403, 2 pls.) has published a very interesting account of some of Onions's experiments during 1913 and 1914. The cape honey-bee, which is very close to the typical form of the species, has been identified by some investigators as *Apis mellifica* var. *kaffra* Lep., by others as subsp. *unicolor* var. *intermissa* Latr. Dissection proves that the workers are peculiar in possessing a well-developed spermatheca. "Careful examination, however, shows the laying workers to contain no spermatozoa, and the development of the sperm-sac must apparently be regarded as merely in some way correlated to the reproductive potentialities of the insect, the organ itself being functionless." According to the experiments carefully reported by Jack, there is every reason to suppose that the workers produce not only males and workers, but even queens parthenogenetically!

17. **Müller, Hermann,** Anwendung der Darwinischen Lehre auf Bienen. *l.c.* 1872, p. 32 *et seq.* He carefully weighed the caterpillar brought into its burrow by an *Ammophila sabulosa* and found it to be six times as heavy as the wasp, whereas the pollen and honey ration provided for each larva by such bees as Diphysis, Colletes and Megachile weighed only twice or thrice as much as the adult insect.

LECTURE IV

1. The curious reader will find attempts at an interpretation of the legend of the gold-digging ants in the following works: **Graf von Veltheim,** Von den goldgrabenden Ameisen und Greiffen der Alten. Helmstädt 1799, 32 pp.; **Keferstein, A.,** Ueber die goldgrabenden Ameisen der Alten. Isis 2, 1825, p. 105–114.

2. As stated in the preface, there will be found an adequate citation of the literature on ants up to 1908 in my book: Ants, Their Structure, Development and Behavior, 1910, p. 578–648. In that work there is also a more detailed treatment of many of the subjects discussed in this and the following lecture. The following authoritative general works, which have appeared in Europe during more recent years, should also be consulted: **Donisthorpe, H. St. J. K.,** British Ants, Their Life History and Classification. Plymouth, Brendon & Son, 1915; **Escherich, K.,** Die Ameise, Schilderung ihrer Lebensweise. 2te Aufl. Braunschweig, Viehweg. 1917; **Forel, A.,** Le Monde Social des Fourmis.

Vols. 1 and 2, Genève, Kundig. 1921 (four more volumes to appear).
3. **Maeterlinck, G.,** J. H. Fabre et Son Œuvre. Ann. Polit. et Litt. Apr. 2, 1911. The passage is translated from the introduction to E. L. Bouvier's book, La Vie Psychique des Insectes. Paris, Flammarion, 1918, p. 3.
4. **Espinas, A.,** Des Sociétés Animales, Étude de Psychologie Comparée. Paris, Germer, Baillière & Co., 1877, p. 213.
5. Most myrmecologists recognize only five subfamilies of ants and regard the Cerapachyinæ as belonging to the Ponerinæ, the Pseudomyrminæ to the Myrmicinæ. It is probable, however, that future systematists will increase the number of subfamilies. I believe that the tribe Leptanillini, which Emery includes among the Dorylinæ, will have to be separated out as a distinct subfamily (Leptanillinæ). Dr. George C. Wheeler finds that the larva of Leptanilla is very aberrant, and the characters of the adult are either quite unlike those of other Dorylinæ or only superficially similar and due to convergence, or similarity of subterranean habits.
6. For fuller treatment of the feeding habits of ants see: **Wheeler, W. M.,** A Study of Some Ant Larvæ, with a Consideration of the Origin and Meaning of the Social Habit among Insects. Proc. Amer. Phil. Soc. 57, 1918, p. 293–343, 12 figs.; **Wheeler, W. M.,** and **Bailey, I. W.,** The Feeding Habits of Pseudomyrmine and Other Ants. Trans. Amer. Phil. Soc. N. S. 22, 1920, p. 235–279, 5 pls., 6 figs.; **Bailey, I. W.,** The Anatomy of Certain Plants from the Belgian Congo, with Special Reference to Myrmecophytism. Bull. Amer. Mus. Nat. Hist. 45, 1922, p. 585–621, 16 pls.
7. See for an account of the habits of driver ants of the Neotropical Region and a citation of the literature my paper: Observations on Army Ants in British Guiana. Proc. Amer. Acad. Arts Sc. 56, 1921, p. 291–328, 10 figs.
8. The latest general account of the honey-ants is given in my paper: Honey Ants, with a Revision of the North American Myrmecocysti. Bull. Amer. Mus. Nat. Hist. 24, 1908, p. 345–397, 28 figs. The following more special and more recent papers may also be cited: **Garrett, A. O.,** Honey Ants in Utah. Science N. S. 32, 1910, p. 342–343; **Leonard, P.,** The Honey Ants of Point Loma, California. Trans. San Diego Soc. Nat. Hist. 1, 1911, p. 85–113, 6 figs.; A Marvel of Motherhood. A Record of Observations on the Founding of a Colony of Honey Ants (*Myrmecocystus mexicanus*). Theosophical Path 6, 1914, p. 225–232; The Living Honey Jars of Lomaland. Raja-Yoga Messenger, July, 1914, p. 19–20, 3 figs.; **Wheeler, W. M.,** Additions to our Knowledge of the Ants of the Genus Myrmecocystus Wesmael.

Psyche 19, 1912, p. 172–181, 1 fig.; The Pleometrosis of Myrmecocystus. *ibid.* 24, 1917, p. 180–182.

9. The following recent literature deals with harvesting ants:
Bouvier, E. L., Sur les Fourmis moissonneuses (*Messor barbara*) des
environs de Royan. 1er Congr. Intern. Ent. 1, Mém. 1911, p. 237–
248; **Doflein, F.,** Mazedonische Ameisen. Beobachtungen über ihre
Lebensweise, Jena, G. Fischer 1920, 8 pls., 10 figs.; Mazedonien, Jena,
G. Fischer, 1921, Chapt. 11; **Emery, C.,** Végétarianisme chez les four-
mis. Arch. Sc. Phys. Nat. Genève (4) 8, 1899, p. 488; Alcune
esperienze sulle formiche granivore. Rend. Accad. Sc. Bologna N. S.
16, 1912, p. 107–117, 1 pl.; **Hunter, W. D.,** Two Destructive Texas
Ants. U. S. Dep. Agr. Circ. No. 148, 1912, 7 pp. (*Atta texana* and
Pogonomyrmex barbatus molefaciens); **Johnston, E. L.,** The Western
Harvesting Ants. Guide to Nature 5, No. 7, 1912, p. 210–214, 4 figs.;
Laloy, L., Les Fourmis moissonneuses. Naturaliste 32, 1910, p. 107–
108 (*Messor*); **Leonard, P.,** The Emmet Harvesters of Point Loma.
The Theosophical Path 2, 1912, p. 215–216, 1 fig.; **Neger, F. W.,** Neue
Beobachtungen an körnersammelnden Ameisen. Biol. Centralbl. 30,
1910, p. 138–150, 3 figs. (*Messor*); **Pierce, W. D.,** The Nest-building
Habits of *Pogonomyrmex barbatus molefaciens* Buckley. Proc. Ent.
Soc. Wash. 12, 1910, p. 97–98; **Turner, C. H.,** The Mound of *Pogo-
nomyrmex badius* Latr. and its Relation to the Breeding Habits of
the Species. Biol. Bull. 17, 1909, p.

10. The following recent literature deals for the most part exclusively with the fungus-growing ants of the tribe Attiini: **Bailey, I. W.,**
Some Relations between Ants and Fungi. Ecology 1, 1920, p. 174–
189, 3 pls.; **Barreto, B. T.,** La Bibijagua y Modos de Combatirlo.
Est. Exper. Agron. Sant. de las Vegas, Cuba, Vol. 42, 1919, p. 9–17,
8 figs. (*Atta insularis*); **Branner, J. C.,** Geologic Work of Ants in
Tropical America. Bull. Geol. Soc. Amer. 21, 1910, p. 449–496, 1
pl., 11 figs.; Repr. in Ann. Rep. Smithson. Inst. Wash. 1911, p.
303–333, 1 pl.; **Bruch, C.,** Costumbres y Nidos de Hormigas. Ann.
Soc. Cient. Argentina 83, 1917, p. 302–316, 11 figs.; Contribucion al
Estudio de las Hormigas de la Provincia de San Luis. Rev. Mus.
La Plata 23, 1916, p. 291–357, 12 pls., 20 figs.; Nidos y Costumbres
de Hormigas. Rev. Soc. Argent. Cienc. Nat. 4, 1919, p. 579–581, 2
figs.; Estudios Mirmecologicos. Rev. Mus. La Plata 26, 1921, p. 175–
211, 6 pls., 15 figs.; Regimen de alimentacion de algunas hormigas
cultivadoras de hongos. Physis 5, 1922, p. 307–311, 3 figs.; **Davis,
W. T.,** The Fungus-Growing Ant on Long Island. Journ. New York
Ent. Soc. 22, 1914, p. 64–65; **Gallardo, A.,** Notas acerca de la Hormiga
Trachymyrmex pruinosus Emery. Ann. Mus. Nac. Hist. Nat. Buenos

Aires 28, 1916, p. 24–252, 4 pls.; Notes Systématiques et Éthologiques sur les Fourmis Attines de la République Argentine. *ibid.* 28, 1916, p. 317–344, 3 figs.; **von Graumitz, C.**, Die Blattschneider-Ameisen Südamerikas. Intern. Ent. Zeitschr. Guben. 7, 1913, p. 233, 240–242; **Ludwig, F.**, Neuere Beobachtungen über pilzezüchtende Insekten. Prometheus 19, 1908, p. 373–374; **Neger, F. W.**, Tiere als Pflanzenzüchter. Kosmos, Stuttgart. 7, 1910, p. 298–301; **Prell, H.**, Biologische Beobachtungen an Termiten und Ameisen. Zool. Anzeig. 38, 1911,

FIG. 114. — Fungus garden of *Cyphomyrmex rimosus* var. *pencosensis,* natural size. (Photograph by Carlos Bruch.)

p. 243–253, 4 figs.; **Spegazzini, C.**, Descripcion de Hongos Mirmecófilos. Dev. Mus. La Plata 26, 1921, p. 166–173, 4 figs.; **Wheeler, W. M.**, Descriptions of Some New Fungus-growing Ants from Texas, with Mr. C. G. Hartman's Observations on Their Habits. Journ. N. Y. Ent. Soc. 19, 1911, p. 245–255, 1 pl.; Two Fungus-growing Ants from Arizona. Psyche 19, 1911, p. 93–101, 2 figs.

11. *Cf.* the classical work on fungus-growing ants: **Moeller, A.**, Die Pilzgärten einiger südamerikanischer Ameisen. Heft 6 of Schimper's Botanische Mittheilungen aus den Tropen 1893, 127 pp., 7 pls.

12. **Spegazzini**, Descripcion de Hongos Mermecófilos *l.c.*

13. A British mycologist (**Farquarson, C. O.**, The Growth of Fungi on the shelters built over Coccids by Cremastogaster Ants. Trans.

FIG. 115. — A portion of the fungus garden of *C. pencosensis* shown in the preceding figure, magnified six diameters. (Photograph by Carlos Bruch.)

FIG. 116. — Another portion of the fungus garden shown in Figs. 114 and 115 and under the same magnification as the latter but showing the Tyridiomyces bromatia more clearly. (Photograph by Carlos Bruch.)

Proc. Ent. Soc. London, 1914, p. XLII–L.) has expressed the opinion that I described Tyridiomyces (The Fungus-growing Ants of North America. Bull. Amer. Mus. Nat. Hist. 23, 1907, p. 669–807, 5 pls., 31 figs.) from asexual spores, but Dr. Carlos Bruch has recently written me that he and Spegazzini have found the same or a very closely allied fungus actively growing in the nests of the Argentinian *Cyphomyrmex rimosus* var. *pencosensis* For. Their beautiful photographs, reproduced in the accompanying Figs. 114 to 116, show that it grows in precisely the manner I described on a substratum of insect excrement. I may add that I examined dozens of nests of *C. rimosus* and its varieties in British Guiana during 1920 and in Panama during 1923 and never failed to find the same fungus actively proliferating and not in the form of asexual spores.

14. **Sampaio de Azevedo, A. G.,** Saúva ou Manhuàara. Monographia, São Paulo 1894; **von Ihering, H.,** Die Anlage neuer Colonien und Pilzgärten bei *Atta sexdens.* Zool. Anzeig. 21, p. 238–245, 1 fig.; **Huber, J.,** Ueber die Koloniegründung bei *Atta sexdens.* Biol. Centralbl. 25, p. 606–619, 625–635, 26 figs. Transl. in Smithson. Report for 1906, 1907, p. 355–367, 5 pls.; A origem das colonias de Saúba (*Atta sexdens*), Bol. Mus. Goeldi 5, 1, p. 223–241; **Goeldi, E. A.,** Beobachtungen über die erste Anlage einer neuen Kolonie von *Atta cephalotes.* C. R. 6ᵐᵉ Congr. Intern. Zool. Berne, 1905, p. 457–458; Myrmecologische Mitteilung das Wachsen des Pilzgärtens bei *Atta cephalotes* betreffend. *ibid.* p. 508–509; **Bruch, C.,** Estudios Mirmecológicos, *l.c.*

15. **Bailey, I. W.,** Some Relations between Ants and Fungi. *l.c.* 1920; The Anatomy of Certain Plants, etc., *l.c.* 1922, p. 607, 608.

LECTURE V

1. **Clarke, J. M.,** Organic Dependence and Disease, Their Origin and Significance. New Haven, Yale Univ. Press, 1921.

2. **Massart, J.,** and **Vandervelde, E.,** Parasitisme Organique et Parasitisme Sociale. Bull. Sc. France Belg. 25, 1893.

3. For the most recent and temperate discussions of the relationships of ants and higher plants (myrmecophytism) see the two following papers: **Bequaert, J.,** Ants in their Diverse Relations to the Plant World. Bull. Amer. Mus. Nat. Hist. 45, 1922, pp. 333–583, 4 pls. (in my Ants of the American Museum Congo Expedition); **Bailey, I. W.,** The Anatomy of Certain Plants from the Belgian Congo, etc.

ibid. l.c.; Notes on Neotropical Ant-Plants I. *Cecropia angulata* sp. nov. Bot. Gazette 74, 1922, p. 369–391, 1 pl., 8 figs. 4. The following contributions to our knowledge of social parasitism among ants have appeared since the publication of my ant book: **Bönner, W.,** Der temporäre soziale Hyperparasitismus von *Lasius fuliginosus* und seine Beziehungen zu *Claviger longicornis* Müll. Zeitschr. wiss. Insektenbiol. 11, 1915, p. 14–20; Ueber die Ursachen der künstlichen Allianzen bei den Ameisen, ein Problem der vergleichenden Psychologie. Zeitschr. wiss. Insektenbiol. 9, 1912, p. 334; **Brun, R.,** Zur Biologie und Psychologie von *Formica rufa* und anderen Ameisen. Biol. Centralbl. 30, 1910, p. 524–528, 529–545; Weitere Beiträge zur Frage der Koloniegründung bei den Ameisen, mit besonderer Berücksichtigung der Phylogenese des sozialen Parasitismus und der Dulosis bei Formica. *ibid.* 32, 1912, p. 154–180, 216–226, 1 fig.; Ueber die Ursachen der künstlichen Allianzen bei den Ameisen. Journ. Psych. u. Neurol. 20, 1913, p. 171–181; **Burrill, A. C.,** How Sanguinary Ants Change at Will the Direction of Column in their Forays. Bull. Wis. Nat. Hist. Soc. 8, 1910, p. 123–131, 2 figs.; **Crawley, W. C.,** Queens of *Lasius umbratus* Nyl. accepted by Colonies of *Lasius niger* L. Ent. Month. Mag. (2) 20, 1909, p. 94–99; Summary of Experiments with fertile females of several species of Ants. Ent. Rec. Journ. Var. 22, 1910, p. 152–156; Workers of *Lasius flavus* (? *umbratus*) among *L. fuliginosus. ibid.* 22, 1910, p. 67–69; *Anergates atratulus* Schenk, a British Ant and the Acceptance of a Female by *Tetramorium cæspitum* L. *ibid.* 24, 1912, p. 218–219; Further Experiments on the Temporary Social Parasitism in Ants of the Genus Lasius Fab., with a Note on *Antennophorus uhlmanni. ibid.* 25, 1913, p. 135–138; Natural Combined Colonies of Ants. *ibid.* 26, 1914; **Crawley, W. C.,** and **Donisthorpe, H. St. J. K.,** The Founding of Colonies by Queen Ants. Trans. 2nd Intern. Congr. Ent. Oxford, p. 11–77; **Davis, W. T.,** Miscellaneous Notes on Collecting in Georgia. Journ. N. Y. Ent. Soc. 19, 1911, p. 216–219 (Incipient *exsectoides-fusca* colony); **Donisthorpe, H. St. J. K.,** *Formica sanguinea* Latr. at Bewdley, with an Account of a Slave-Raid and Description of Two Gynandromorphs, etc. Zoologist, 1909, p. 463–466; On the Founding of Nests by Ants and a Few Notes on Myrmecophiles. Ent. Rec. Journ. Var. 22, 1910, p. 83–84; Some Experiments with Ants' Nests. Trans. Ent. Soc. London, 1910, p. 142–150; On Some Remarkable Associations between Ants of Different Species. Rep. Lancas. Ches. Ent. Soc. 36, 1912, 19 pp.; Nest of *Lasius fuliginosus.* Trans. Ent. Soc. London, 1914, p. XVIII–XIX; British Ants, Their Life History and Classification, 1915, *l.c.;* Marriage-flights of Donisthorpea species on

DOCUMENTARY APPENDIX   341

August 8, etc. Ent. Rec. Journ. Var. 27, 1915, p. 207; Myrmecophilous Notes for 1916, *ibid.* 28, 1916, p. 2–3; for 1916, *ibid.* 29, 1917, p. 32–33; for 1917, *ibid.* 30, 1918, p. 13; for 1918, *ibid.* 31, 1919, p. 3–4; The Colony Founding of *Acanthomyops* (*Dendrolasius*) *fuliginosus* Latr. Biol. Bull. 42, 1922, p. 173–184; **Donisthorpe, H. St. J. K.** and **Crawley, W. C.**, Experiments on the Formation of Colonies by *Lasius fuliginosus* Females. Trans. Ent. Soc. London 1911, p. 664–672; Further Observations on Temporary Social Parasitism and Slavery in Ants. *ibid.* 1911, p. 175–183; **Emery, C.**, Intorno all' origine delle Formiche dulotiche parassitiche e mirmecofile. Rend. Accad. Sc. Bologna N. S. 13, 1909, p. 36–50; Beobachtungen und Versuche an *Polyergus rufescens*. Biol. Centralbl. 31, 1911, p. 625–642; Einiges über die Ernährung der Ameisenlarven und die Entwicklung des temporären Parasitismus bei Formica. Deutsch. Ent. Nat. Biblioth. 2, 1911, p. 4–6; Ulteriori osservazioni ed esperienze sulla Formica amazzone. Rend. Accad. Sc. Bologna N. S. 15, 1911, p. 60–75; Ueber die Abstammung der europäischen arbeiterinnenlosen Ameise " Anergates." Biol. Centralbl. 33, 1913, p. 258–260; Histoire d'une Société expérimentale de *Polyergus rufescens*. Rev. Suisse Zool. 23, 1915, p. 385–400, 2 figs.; Histoire d'une Société expérimentale de Fourmi amazone. Act. Soc. Helvet. Sc. Nat. 97ᵐᵉ sess. 2, 1915–16, p. 272–273; **Emmelius, C.**, Beiträge zur Biologie einiger Ameisenarten. Biol. Centralbl. 39, 1919, p. 303–311; **Forel, A.**, Fondation des fourmilière de *Formica sanguinea*. Arch. Sc. Phys. Nat. (4) 28, 1909, 2 pp.; Une colonie polycladique de *Formica sanguinea* sans esclaves dans le canton de Vaud. 1ᵉʳ Congr. Intern. Ent. Mém. 1, p. 101–104; **Krausse, A.**, *Formica fusca fusca*-Königin bei *Formica rufa pratensis*-Arbeiterinnen im künstlichen Nest. Biol. Centralbl. 41, 1921, p. 523–527; **Kutter, H.**, Ein weiterer Beitrag zur Frage der sozial-parasitischen Koloniegründung von *F. rufa* L. Zugleich ein Beitrag zur Biologie von *F. cinerea*. Zeitschr. wiss. Insektenbiol. 9, 1913, p. 193–196; Zur Biologie von *Formica rufa* und *Formica fusca* i. sp. Biol. Centralbl. 33, 1913, p. 703–707; *Strongylognathus Huberi* Forel r. *alpinus* Wh. eine sklavenraubende Ameise. *ibid.* 40, 1920, p. 528–538, 1 fig.; **Lomnicki, J.**, Sur la micrandrie chez la fourmi sanguine (Polish). Kosmos. Bull. Soc. Polon. Natural. 1921, p. 98–99; Ueber den Anfang der Kolonien der glänzendschwarzen Holzameise. Ent. Anzeig. 2, 2 pp. (*Lasius fuliginosus*); **Mann, W. M.**, Parabiosis in Brazilian Ants. Psyche 19, 1912, p. 36–41; **McColloch, J. W.** and **Hayes, W. P.**, A Preliminary Report on the Life Economy of *Solenopsis molesta* Say. Journ. Econ. Ent. 9, 1916, p. 23–38, 1 pl., 1 fig.; **Piéron, H.**, La génèse des instincts esclavagistes et parasitaires chez les Fourmis. Rev. Gén.

Sc. 21, 1910, p. 726–736, 769–779; On the Origin of Slavery and Parasitism in Ants. Nature 85, 1910, p. 351–352; **Reichensperger, A.,** Beobachtungen an Ameisen. Biol. Centralbl. 31, 1911, p. 596–605; **Rüschkamp, F.,** Eine neue natürliche *rufa fusca*-Adoptionskolonie. *Ibid.* 32, 1912, p. 213–216; Eine dreifach gemischte natürliche Kolonie (*Formica sanguinea-fusca-pratensis*). *ibid.* 33, 1913, p. 668–672; **Santschi, F.,** Une nouvelle fourmi parasite. Bull. Soc. Hist. Nat. Afr. Nord. 5, 1913, p. 229–230 (*Wheeleriella adulatrix* n. sp.); Fourmis du Genre Bothriomyrmex Emery. Rev. Zool. Afr. 7, 1920, p. 201–224, 2 pls., 2 figs.; **Stumper, R.,** *Formicoxenus nitidulus* Nyl. 1. Beitrag. Biol. Centralbl. 38, 1918, p. 160–179, 14 figs.; Psychobiologische Beobachtungen und Analysen an Ameisen. *ibid.* 38, 1918, p. 345–354, 2 figs.; Pages Myrmécologiques. 1. *Formicoxenus nitidulus* Nyl. Bull. Soc. Natural. Luxemb. 1918 (?), 10 pp., 3 figs.; Zur Kenntnis des Polymorphismus der Formiciden (*Formicoxenus nitidulus* Nyl.). Ver. Ges. Luxemb. Naturf. 1919, 7 pp., 8 figs.; Zur Ontogenese der Ameisenkolonien. 1. Beitrag. Arch. Naturg. 83, 1919, p. 1–10, 1 fig.; 2. Beitrag. *ibid.*, p. 137–141; Zur Kolonie-Grundung von *Lasius fuliginosus*. *ibid.* 85, 1920, p. 189; Études sur les Fourmis III. Recherches sur l'Éthologie du *Formicoxenus nitidulus* Nyl. Bull. Soc. Ent. Belg. 7, 1921, p. 90–97, 1 fig.; **Szabó, J.,** Magyarország rabszolgatartó es élösködö hangyái. Allatt. Közlem. Kot. 13, 1914, p. 93–105, 7 figs.; Ungarns sklavenhaltende und parasitische Ameisen. *ibid.* p. 149–150; **Tanquary, M. C.,** Experiments on the Adoption of Lasius, Formica and Polyergus Queens by Colonies of Alien Species. Biol. Bull. 20, 1911, p. 281–308; **Torka, V.,** Raubzug von *Polyergus rufescens* Latr. Deutsch. Ent. Zeitschr. 1914, p. 645–646; **Viehmeyer, H.,** Ontogenetische und phylogenetische Betrachtungen über die parasitische Koloniegründung von *Formica sanguinea*. Biol. Centralbl. 30, 1910, p. 569–580; Ueber eine erst in den letzten Jahren in Sachsen aufgefundene Ameise: *Harpagoxenus sublevis* (Nyl.) Iris. 24 Korrespondenzbl., 1910, p. 40; Morphologie und Phylogenie von *Formica sanguinea*. Zool. Anzeig. 37, 1911, p. 427–441; Zur sächsischen Ameisenfauna. Abh. Naturw. Ges. Isis. Dresden. 1915, p. 61–64; **Voss, F.,** Ueber den sozialen Parasitismus der Ameisen. 60/61. Jahresb. Nat. Ges. Hannover. –2/4 Jahresb. niederächs. Zool. Ver. 1912, p. X–XI; **Wasmann, E.,** Zur Geschichte der Sklaverei und des sozialen Parasitismus bei den Ameisen. Naturw. Wochenschr. 27, 1909, p. 401–407, 5 figs.; Ueber gemischte Kolonien von Lasius-Arten. Zool. Anzeig. 35, 1909, p. 129–141; Ueber den Ursprung des sozialen Parasitismus, der Sklaverei und der Myrmekophilie bei den Ameisen. Biol. Centralbl. 29, 1909, p. 587–604, 619–637, 651–663, 683–703, 2 figs.; Nachträge zum

sozialen Parasitismus und der Sklaverei bei den Ameisen. *ibid.* 30,
1910, p. 453–464, 475–496; 515–524; Eine neue natürliche *rufa-fusca*-
Adoptions Kolonie. *ibid.* 32, 1912, p. 213–216; Ein neuer Fall zur
Geschichte der Sklaverei bei den Ameisen. Verh. Ges. deutsch. Nat.
Aertzte Vers. 84, Tl. 2, Hälfte 1, 1913, p. 264–268; Nachschrift (zu
Rüschkamp 1913) Ueber *pratensis* als Sklaven von *sanguinea. ibid.*
33, 1913, p. 672–675; *Anergatides Kohli,* eine neue arbeiterlose
Schmarotzerameise vom oberen Congo. Ent. Mitt. 4, 1915, p. 279–
288, 2 pls.; Das Gesellschaftsleben der Ameisen. Münster i. W.
Aschendorfsche Verlagsbuch. Band I, pp. XX + 413, 7 pls., 16 figs.;
Wheeler, W. M., Observations on Some European Ants. Journ. N. Y.
Ent. Soc. 17, 1909, p. 172–187, 2 figs.; Notes on Some Slave-Raids
of the Western Amazon Ant (*Polyergus breviceps* Emery). *ibid.* 24,
1916, p. 107–118; The Temporary Social Parasitism of *Lasius subum-
bratus* Viereck. Psyche 24, 1917, p. 167–176; The Pleometrosis of
Myrmecocystus. *ibid.,* p. 180–182; The Parasitic Aculeata, A Study
in Evolution. 1919, *l.c.;* A New Case of Parabiosis and the "Ant
Gardens" of British Guiana. Ecology 2, 1921, p. 89–103, 3 figs.;
Neotropical Ants of the Genera Carebara, Tranopelta and Tranopel-
toides, New Genus. Amer. Mus. Novitates 48, 1922, p. 1–14, 3 figs.;
Yano, M., A New Slave-making Ant from Japan. Psyche 18, 1911,
p. 110–112, 1 fig. (*Polyergus rufescens samurai* n. subsp.).

5. For lists of the other known parasitic ants and their hosts see
my paper: The Parasitic Aculeata, etc., 1919 *l.c.*

6. The literature published on myrmecophily since 1908 is too
extensive for citation in this place. I therefore confine myself to a
few works of special interest in connection with the limited field
embraced in the lecture.

7. For a more extended account of the habits of *Metopina pachycondylæ* see my paper: An extraordinary Ant Guest. Amer. Nat. 35,
1901, p. 1017–1016, 2 figs.

8. **Janet, C.,** Sur le *Lepismina polypoda* et sur ses rapports avec
les Fourmis. Bull. Soc. Ent. France 65, 1896, p. 131; Sur les Rapports des Lépismides myrmécophiles avec les Fourmis. C. R. Acad.
Sc. Paris, 122, 1896, p. 799, 1 fig.

9. See especially **Janet, C.,** Sur les rapports de l'*Antennophorus
Uhlmanni* Haller avec le *Lasius mixtus* Nyl. C. R. Acad. Sc. Paris
124, 1897, p. 583–585, 1 fig.; Sur le *Lasius mixtus,* l'*Antennophorus
uhlmanni,* etc. Limoges, 1897, 62 pp., 16 figs.; **Karawaiew, W.,** *Antennophorus uhlmanni* Hall. und seine biologischen Benziehungen zu *Lasius
fuliginosus* und zu andren Ameisen. Mém. Soc. Nat. Kiew (1903) 19,
1905, p. 193–241, 1 pl., 2 figs.; Weitere Beobachtungen über Arten der

Gattung Antennophorus. *ibid.* 20, 1906, p. 209–229 (Russian with German résumé); **Wheeler, W. M.,** Two New Myrmecophilous Mites of the Genus Antennophorus. Psyche 17, 1910, p. 1–6, 2 pls. 10. **Wasmann** (Die Gastpflege der Ameisen, ihre biologischen und philosophischen Probleme. Heft 4 of J. Schaxel's Abhandlungen zur theoretischen Biologie 1920, pp. XVII + 176, 2 pls.) has recently published an elaborate résumé of his 35 years of investigation of Lomechusini and other myrmecophiles, largely as a criticism of and counterblast to my paper on the study of ant-larvæ (1918). As a study in Jesuit psychology the work may be recommended to biologists who can spare the time for its perusal from more important occupations. He has to admit the facts cited in my paper and much of my interpretation of them, but by adroit perversion of my statements, hairsplitting definitions and subtleties and by the production of voluminous smoke-screens of Thomistic argumentation he seeks to conceal the real scientific weakness of his contentions. The paper closes, of course, with an attack on the monistic and a defence of the theistic " Weltanschauung " and of those who have developed not only beyond the philosophy of St. Thomas but also that of the " Aufklärung." The geneticists will, no doubt, be pleased to learn that the genes are "instrumental *causæ secundæ* for the whole of the orderly organic-psychic development of organisms " and that the Divine Wisdom and Omnipotence are still functioning as the " *causa prima*." Wasmann's contribution is valuable chiefly because it contains a good résumé of his previous work and speculations with bibliographies of the 233 books and pamphlets which he had published up to 1918 and of those of other authors who have written on myrmecophiles.

11.   That the trichomal gland secretions may not only attract but in exceptional cases even narcotize the ant that imbibes them, is shown by Jacobson's important observations on *Ptilocerus ochraceus,* a peculiar Javanese bug which possesses a gland with golden trichomes in the midventral line of the second abdominal segment (see **Kirkaldy, G. W.,** Some Remarks on the Reduviid Subfamily Holoptilinæ, and on the Species *Ptilocerus ochraceus* Montandon. Tijdschr. Ent. 54, 1911, p. 170–174, 1 pl., and **Jacobson, E.,** Biological Notes on the Hemipteron *Ptilocerus ochraceus. ibid.* p. 175–179). Jacobson found this insect frequenting the trails of a common East Indian ant, *Dolichoderus bituberculatus* on the rafters of houses, and describes the behavior of both insects as follows: " The way in which the bugs proceed to entice the ants is as follows. They take up a position in an ant-path or ants find out the abodes of the bugs, and attracted by their secretions visit them in great numbers. On the approach of an ant of the species

Dolichoderus bituberculatus the bug is at once on the alert; it raises halfway the front of the body, so as to put the trichome in evidence. As far as my observations go the bugs only show a liking for *Dolichoderus bituberculatus;* several other species of ants, e.g. *Cremato-gaster difformis* Smith and others which were brought together with them, were not accepted; on the contrary, on the approach of such a stranger, the bug inclined its body forwards, pressing down its head; the reverse therefore of the inviting attitude taken up towards *Dolichoderus bituberculatus.* In meeting the latter the bug lifts up its front legs, folding them in such a manner that the tarsi nearly meet below the head. The ant at once proceeds to lick the trichome, pulling all the while with its mandibles at the tuft of hairs, as if milking the creature, and by this manipulation the body of the bug is continually moved up and down. At this stage of the proceedings the bug does not yet attack the ant; it only takes the head and thorax of its victim between its front legs, as if to make sure of it; very often the point of the bug's beak is put behind the ant's head, where this is joined to the body, without, however, doing any injury to the ant. It is surprising to see how the bug can restrain its murderous intention as if it was knowing that the right moment had not yet arrived. After the ant has indulged in licking the tuft of hair for some minutes the exudation commences to exercise its paralyzing effect. That this is only brought about by the substance which the ants extract from the trichome, and not by some thrust from the bug, is proved by the fact, that a great number of ants, after having licked for some time the secretion from the trichome, leave the bug to retire to some distance. But very soon they are overtaken by the paralysis, even if they have not been touched at all by the bug's proboscis. In this way a much larger number of ants is destroyed than actually serves as food to the bugs, and one must wonder at the great prolificacy of the ants, which enables them to stand such a heavy draft on the population of one community. As soon as the ant shows signs of paralysis by curling itself up and drawing in its legs, the bug at once seizes it with its front legs, and very soon it is pierced and sucked dry. The chitinized parts of the ant's body seem to be too hard for the bug to penetrate, and it therefore attacks the joints of the armour. The neck, the different sutures on the thorax and especially the base of the antennæ are chosen as points of attack. Nymphs and adults of the bug act in exactly the same manner to lure the ants to their destruction, after having rendered them helpless by treating them to a tempting delicacy."

These interesting observations by a very competent naturalist are important in connection with the probable significance of the trichomes in some other ant-guests and predators, and especially in connection with Wasmann's lucubrations in regard to Lomechusa. To assume that *D. tuberculatus* has acquired through "amical selection" a peculiar Ptilocerus-licking instinct, would be absurd. We are obviously confronted with a flagrant example of appetite perversion.

12. **Hobhouse, L. T.,** Mind in Evolution. London, Macmillan 1915, p. 97.

LECTURE VI

1. It is very doubtful whether the peculiar insects belonging to the order Zoraptera are social or even subsocial. Wingless specimens of a species (*Zorotypus hubbardi*) were first encountered by H. G. Hubbard near Hawk Creek, Florida, in 1895, in a nest of *Reticulitermes flavipes*. They were taken in 1918 at Miami in the same state by T. E. Snyder in a nest of *Prorhinotermes simplex* but were found also in portions of logs where there were no termites (see **Banks** and **Snyder,** A Revision of the Nearctic Termites, Bull. 108, U. S. Nat. Mus. 1920, p. 119). Hubbard regarded his specimens as wingless Psocids and it was not till 1913 that Silvestri created the new order Zoraptera for an East Indian species (Descrizione di un nuovo ordine di insetti. Boll. Lab. Zool. Portici, 7, 1913, p. 193–209). More recently **Caudell, A. N.** (Zoraptera not an Apterous Order. Proc. Ent. Soc. Wash. 22, 1920, p. 84–97, 1 pl.), has shown that the Florida species may have wings, and **Crampton, G. C.** (Some Anatomical Details of the Remarkable Winged Zorapteron, *Zorotypus hubbardi* Caudell, with Notes on its Relationships. *ibid.* 22, 1920, p. 98–106, 1 pl., 1 fig.), has discussed their affinities to the termites and allied orders. Caudell has also described a species from Hawaii (*Zorotypus swezeyi*, A New Species of the Order Zoraptera from Hawaii. Trans. Amer. Ent. Soc. 48, 1922, p. 133–135), and during the summer of 1920 Alfred Emerson found a specimen of an undescribed species in a termite nest in British Guiana. The species of the order are therefore widely distributed. In his paper of 1920 Caudell makes the following statement: "Both species of Zorotypus are social insects, occurring in colonies of various sizes. They probably occur near Termites, but are not actually mingled with them and are probably never really inquilinous with them as was at first thought probable, due to their usual proximity to white ants and their frequent occupancy of their galleries." Apparently the only reasons for supposing that the Zoraptera are social is the fact that they have winged and wingless forms and that the

latter may lose their wings like male and female termites. It seems more probable that Zorotypus is merely gregarious like many Psocids.
2. The account of the habits of the Dermaptera is drawn from the following literature: **Bennett, C. B.**, Earwigs (*Anisolabia maritima*). Psyche 11, 1904, p. 47–53, 2 figs.; **De Geer, C.**, Mémoires pour servir a l'Histoire des Insectes 3, 1773, Stockholm, P. Hesselberg, p. 548; **Frisch, J. L.**, Beschreibung von allerley Insecten in Deutschland, etc. Berlin, C. Gottl. 8 Theil, 1730, p. 32; **Jones, D. W.**, The European Earwig and its Control. Bull. 566, U. S. Dep. Agric. 1917, 12 pp., 8 figs.; **Morse, A. P.**, Manual of the Orthoptera of New England. Proc. Boston Soc. Nat. Hist. 35, 1920, p. 197–556, 20 pls.
3. For the taxonomy, distribution and habits of the Embidaria the following literature may be consulted: **Cockerell, T. D. A.**, Descriptions of Tertiary Insects II. Amer. Journ. Sc. (4) 25, p. 227–232, fig. (*Embia florissantensis*); **Enderlein, G.**, Uber die Morphologie, Gruppierung und systematische Stellung der Corrodentien. Zool. Anzeig. 26, 1903, p. 423–437, 4 figs.; Die Klassifikation der Embiidinen, nebst morphologischen und physiologischen Bemerkungen, besonders über das Spinnen derselben. *ibid.* 35, 1909, p. 166–191, 3 figs.; **Friederichs, K.**, Zur Biologie der Embiidinen. Mitt. zool. Mus. Berlin 3, 1906, p. 215–239, 19 figs.; Zur Systematik der Embiiden. Verh. zool.-bot. Ges. Wien 57, 1907, p. 270–275; **Grassi, B.**, and **Sandias, A.**, Contribuzione allo Studio delle Embidine, p. 133–150, 4 pl., 11 figs. Appendix to their Constituzione e sviluppo della Società dei Termitidi (*vide infra*); **Hagen, H.**, Monograph of the Embiidina. Canad. Ent. 17, 1885, p. 141–155, 171–178; **Imms, A. D.**, Contributions to a Knowledge of the Structure and Biology of Some Indian Insects. II. On *Embia major* sp. nov. from the Himalayas. Trans. Linn. Soc. London Zool. (2) 11, 1913, p. 167–195, 3 pls., 6 figs.; **Krauss, H. A.**, Monographie der Embien. Zoologica 23, Heft 60, 1911, 78 pp., 5 pls., 7 figs.; **Kusnezow, N. J.**, Observations on *Embia taurea* Kusnez. (1903) from the Southern Coast of Crimea (Russian, with English résumé). Hor. Soc. Ent. Ross. 37, 1904; **Lucas, H.**, Histoire Naturelle des Animaux articulés de l'Algérie. Explor. Sc. d'Algér. Zool. 3, 1849, p. 111–114, pl. 3, fig. 2 (*Embia mauritanica*); Quelques Remarques sur la propriété que possède la larve de l'*Embia Mauritanica* de secréter une matière soyeuse destinée a construire des fourreaux dans lesquels elle subit ses divers changements de peau. Ann. Soc. Ent. France (5) 10. 1859, p. 441–444; **MacLachlan, R.**, On the Nymph-stage of the Embidæ, with Notes on the Habits of the Family, etc. Proc. Linn. Soc. London 13, 1877, p. 373–384, 1 pl.; **Melander, A, L.**, Two New Embiidæ. Biol. Bull. 3, 1902, p. 16–26, 4 figs.; Notes on the Structure and

Development of *Embia texana*. Biol. Bull. 4, 1903, p. 99–118, figs.;
Michael, W. H., A New Danger for Orchid Growers. Gard. Chron.
6, 1876, p. 845, fig. (*Oligotoma michæli*); **Perkins, R. C. L.**, Notes
on *Oligotoma insularis* McLachl. (Embiidæ), and its immature Conditions. Ent. Month. Mag. (2) 8, 1897, p. 56–58; **Pictet, F. J.**, Traité
de paléontologie ou histoire naturelle des animaux fossiles. 2ᵉ edit.
Paris, Vol. 2, p. 370, pl. 40, fig. 28 (*Embia antiqua*); **Rimsky-
Korsakow, M.**, Beitrag zur Kenntnis der Embiiden. Zool. Anzeig. 29,
1905, p. 433–442, 5 figs.; Ueber das Spinnen der Embiiden. *ibid.*
36, 1910, p. 153–156, 2 figs.; Regenerationserscheinungen bei Embiiden.
Verh. 3. Intern. Zool. Congr. Graz. 1910, p. 609–620; Ueber den
Bau und die Entwicklung des Spinnapparates bei Embien. Zeitschr.
wiss. Zool. 108, 1914, p. 499–519, 2 pls., 1 fig.; **de Saussure, H.**, Note
sur la Tribu des Embiens. Mitt. Schweiz. Ent. Ges. 9, 1896, p. 339–
355, 1 pl.; Two Embidæ from Trinidad. Journ. Trinidad Club 2,
1896, p. 292–294; **Verhoeff, K. W.**, Zur vergleichenden Morphologie
und Systematik der Embiiden. Act. Acad. Leop. Halle 82, 1904,
p. 145–204, 4 pls.; **Westwood, J. O.**, Characters of Embia, a Genus of
Insects allied to the White-Ants (Termites), with Descriptions of the
Species of which it is composed. Trans. Linn. Soc. London 17, 1837,
p. 369–375, 1 pl.; **Woodmason, J. A.**, Contribution to our Knowledge
of the Embiidæ, a Family of Orthopterous Insects. Proc. Zool. Soc.
London, 1883, p. 628–634, 1 pl.

4. The following note in Krauss's monograph (*l.c.* 1911, p. 18) is
also interesting in this connection. "The eggs are deposited in the
nest, singly or in clusters, often as many as several dozen together,
and in their immediate neighborhood gnawed pieces of plants, detritus
and especially bitten-off stamens are accumulated as food for the
young, as has been observed especially in the case of *Embia mauritanica* by Vosseler. As early as 1894 he called my attention to this
form of parental solicitude which in the meantime has been observed
by Friederichs in *Haploëmbia solieri*. The eggs and young larvæ are
cared for by the mother in precisely the same manner as in the case
of the Forficulids."

5. Among the following list of Termite literature, selected as important for a general survey of the order, the works of Escherich
(1909, 1911), Hegh (1922) and Holmgren (1909–1913) are the most
comprehensive and contain very full bibliographies: **Andrews, E. A.**,
Observations on Termites in Jamaica. Journ. Anim. Behav. 1, 1911,
p. 193–228; **Banks, N.**, and **Snyder, T. E.**, Revision of the Nearctic
Termites with Notes on Biology and Geographic Distribution. Bull.
108, U. S. Nat. Mus. 1920, 228 pp., 35 pls., 70 figs.; **Bugnion, E.**,

Observations relatives a l'Industrie des termites. Ann. Soc. Ent.
France 79, 1910, p. 129–144; Différentiation des Castes chez les Termites. Bull. Soc. Ent. France 1913, p. 213–218; Les Termites
de Ceylon avec quelques Indications sur la distribution géo-
graphique de ces Insectes. Le Globe 52, 1913, 36 pp., 8 pls.; **Desneux, J.,**
Isoptera, Fam. Termitidæ. Fasc. 25, Wytsman's Genera Insectorum
1904, 52 pp., 2 pls., 10 figs.; **Drummond, H.,** On the Termite as the
tropical Analogue of the Earthworm (Central Africa). Proc. Roy.
Soc. Edinb. 13, 1886, p. 137–146; Tropical Africa, New York, Scribner
& Welford 1889, Chapt. VI; **Dudley, P. H.,** Observations on the Ter-
mites or White Ants of the Isthmus of Panama. Trans. N. Y. Acad.
Sc. 8, 1888–1889, p. 85–114; **Escherich, K.,** Die Termiten oder weissen
Ameisen. Leipzig, W. Klinkardt 1909, 198 pp., 51 figs.; Termitenleben
auf Ceylon. Jena, G. Fischer 1911, pp. XXXII + 262, 3 pls., 67 figs.;
Froggatt, W. W., The White Ants' City, a Natural History Study
Agric. Gaz. N. S. Wales 14, 1903, p. 729; White Ants (Termitidæ).
Misc. Publ. No. 874, Dept. Agric. N. S. Wales 1905, 47 pp., 5 pls.,
9 figs.; **Grassi, B.,** and **Sandias, A.,** Constituzione e Sviluppo della So-
cieta dei Termitidi. Atti Accad. Gioenia (4) 6–7, 1893–94, 151 pp.,
5 pls.; **Hagen, H.,** Monographie der Termiten. Linn. Ent. Stett Ent.
Ver. 10, 1855; 12, 1858; 14, 1860; **Haviland, G. D.,** Observations on
Termites, with Description of New Species. Journ. Linn. Soc. London
26, 1898, p. 338–442; Observations on Termites or White Ants. Ann.
Rep. Smithson. Inst. for 1901, 1902, p. 667–678, 4 pls.; **Heath, H.,** The
Habits of California Termites. Biol. Bull. 4, 1903, p. 47–63; **Hegh, E.,**
Les Termites. Partie Générale, Brussels, Imp. Indust. Finan. 1922,
756 pp., 460 figs., 1 map; **Holmgren, N.,** Studien über südamerikanische Termiten. Zool. Jahrb. Syst. 23, 1906, p. 371–676; Termiten-
studien. 4 parts. K. Svensk. Vetensk. Handl. 44, 1909, 215 pp., 3 pls.,
76 figs.; 46, 1911, 86 pp., 6 pls., 6 figs.; 48, 1912, 166 pp. 4 pls.,
88 figs.; 50, 1913, 276 pp., 8 pls., 14 figs.; **Imms, A. D.,** On the
Structure and Biology of Archotermopsis, together with Descriptions
of New Species of Intestinal Protozoa and General Observations on
the "Isoptera." Phil. Trans. Roy. Soc. London (B) 209, 1919, p. 75–
180, 7 pls.; **König, J. G.,** Naturgeschichte der sogenannten weissen
Ameisen. Beschr. Ber. Ges. Naturf. Freunde Berlin 4, 1779, p. 1–28;
Lespès, C., Recherches sur l'organisation et les mœurs du Termite
lucifuge Ann. Sc. Nat. Zool. (4) 5, 1856, p. 227–282, 3 pls.; **Müller, F.,**
Beiträge zur Kenntnis der Termiten. Jen. Zeitschr. 7, 1873, 2 pls.;
von Rosen, K., Die fossilen Termiten: eine kurze Zusammenfassung der
bis jetzt bekannten Funde. Trans. 2nd Intern. Congr. Ent., Vol. 2,
p. 318–334, 6 pls.; **Silvestri, F.,** Contribuzione alla Conoscenza dei

Termitide e Termitofili dell' America Meridionale. Redia 1, 1903, p. 1–235, 6 pls.; Contribuzione alla Conoscenza dei Termitidi e Termitofili dell Africa Occidentale. Boll. Lab. Zool. Portici 9, 1914, p. 1–146, 1 pl., 84 figs.; 12, 1918, p. 287–346, 47 figs.; 14, 1920, p. 265–319, 32 figs.; **Sjöstedt, Y.**, Monographie der Termiten Afrikas. K. Svensk. Vet. Akad. Handl. 34, 1900, 236 pp., 9 pls.; Nachtrag, *ibid.* 38, 1904, pp. 120, 4 pls.; **Smeathman, H.**, Some Account of the Termites which are found in Africa and Other Hot Climates. Phil. Trans. Roy. Soc. 71, 1781, p. 139–192, 3 pls.; **Snyder, T. E.**, The Biology of the Termites of the Eastern United States with Preventive and Remedial Measures. Bur. Ent. U. S. Dep. Agric., Bull. 94, 1915, 85 pp., 17 pls., 14 figs.; **Thompson, Miss C. B.**, The Brain and the Frontal Gland of the Castes of the White Ant, *Leucotermes flavipes* Koll. Journ. Comp. Neurol. 26, 1916, p. 553–603, 26 figs.; Origin of the Castes of the Common Termite, *Leucotermes flavipes* Kol. Journ. Morph. 30, 1917, p. 83–152, 42 figs.; The Development of the Castes of Nine Genera and Thirteen Species of Termites. Biol. Bull. 36, 1919, p. 379–398; The Castes of Termopsis. Journ. Morph. 36, 1922, p. 495–534, 2 pls., 9 figs.; **Thompson, Miss C. B.**, and **Snyder, T. E.**, The Question of the Phylogenetic Origin of Termite Castes. Biol. Bull. 36, 1919, p. 115–129, 5 figs.; **Uichanco, L.**, General Facts in the Biology of Philippine Mound-building Termites. Philip. Journ. Sc. 15, 1919, p. 59–64, 4 pls.

6. **Darwin, C.**, The Origin of Species. Chapt. VIII On Neuter and Sterile Insects.

7. A very comprehensive account of the nests of termites with a great many illustrations is given by **Hegh**, Les Termites *l.c.* 1922.

8. The following list comprises most of the works containing accounts of the fungus-growing habits of termites: **Bequaert, J.**, Notes biologiques sur quelques fourmis et termites du Congo Belge. Rev. Zool. Afr. 2, 1913, p. 396; **Brown, W. H.**, The Fungi Cultivated by Termites in the Vicinity of Manila and Los Baños. Philip. Journ. Sc. (C.) Botany 13, 1918, p. 223–231, 2 pls.; **Bugnion, E.**, L'imago du *Coptotermes flavus*. Larves portant des Rudiments d'Ailes prothoraciques. Mém. Soc. Zool. France 24, 1911, p. 97–106, 2 pls., 2 figs.; Les Mœurs des termites champignonnistes de Ceylan. Bibl. Univ. Lausanne, 1913, p. 552–583, 1 pl.; La Biologie des termites de Ceylan. Bull. Mus. d'Hist. Nat. 1914, 38 pp., 8 pls.; **Doflein, F.**, Die Pilzkulturen der Termiten. Verh. deutsch. Zool. Ges. 1905, p. 140–149. Transl. in Spolia Zeylandica 3, 1906, p. 203–209; **Escherich, K.**, Eine Ferienreise nach Erythrea. Leipzig, Quelle u. Mayer 1908; Die Termiten oder weissen Ameisen, *l.c.* 1909; Die pilzzüchtenden Termiten. Biol. Centralbl. 29, 1909, p. 16–27; Termitenleben auf Ceylon, *l.c.* 1911; **Fletcher, T. B.,**

Some Indian Insects and Other Animals of Importance considered especially from an Economic Point of View. Madras, Government Press 1914; **Fuller, C.**, Observations on Some South African Termites. Ann. Natal Mus. 3, 1915, p. 329–504, 11 pls., 16 figs.; The Termites of South Africa, being a Preliminary Notice. S. Afric. Journ. Nat. Hist. Pretoria 1921, p. 14–52; Notes on White Ants. On the Behavior of True Ants towards White Ants. Bull. S. Afr. Biol. Soc. (1) 1, 1918, p. 16–20; White Ant Notes. Journ. Dep. Agric. Un. S. Afr. Pretoria 2 and 3, No. 5, p. 462–466; No. 2, p. 142–147; **Haviland, G. D.**, Observations on Termites, *l.c.* 1898; **Hegh, E.**, Les Termites, *l.c.* 1922; **Holmgren, N.**, Termites from British India (Bombay) collected by Dr. J. Assmuth S. J. Journ. Bombay Nat. Hist. Soc. 1912, p. 774–793, 4 pls.; **Holtermann, C.**, Mykologische Untersuchungen aus den Tropen. 1898, p. 107, 1 pl.; Pilzanbauende Termiten, in Bot. Untersuch, S. Schwendener, 1899, p. 411–420; Fungus Cultures in the Tropics. Ann. Roy. Bot. Gard. Peradenyia 1, 1901, p. 27–37; **Jumelle, H.**, and **Perrier de la Bathie, H.**, Les Termites champignonnistes de Madagascar. C. R. Acad. Sc. Paris 144, p. 1449; Les Champignons des termitières de Madagascar. *ibid.* 145, p. 274; Termites champignonnistes et champignons des termitières de Madagascar. Rev. Gén. Botan. Paris 22, 1910, p. 30 *et seq.;* **Karawaiew, W.**, Nachträge zum vorläufigen Bericht über die Reise auf der Insel Java (Russian) Kiew 1901; The Termite Fungus *Agaricus rajap* Holt. Mém. Soc. Kiew 17, 1902, p. 298–303; **Knuth, P.**, Termiten und ihre Pilzgärten. Ill. Zeitschr. Ent. 4, 1899, p. 257–259; **König, J. G.**, Naturgeschichte der sogenannten weissen Ameisen. *l.c.* 1779; **Morstätt, H.**, Ostafrikanische Termiten I. Allgemeines über Termiten. Der Pflanzer, Amani 9, 1913, p. 130; Die Nataltermite und andere Arten an Kautschukbäumen. *ibid.* p. 443; Ueber einige Ergebnisse der Termitenforschung. Biol. Zentralbl. 40, 1920, p. 415–427; Ueber Pilzgärten bei Termiten. Ent. Mitteil. 11, 1922, p. 94–99; Zur ständischen Gliederung und Ernährungsbiologie der Termiten. Ent. Mitteil. 11, 1922, p. 9–16; **Petch, T.**, The Fungi of Certain Termite Nests. Ann. Roy Bot. Gard. Peradenyia 3, 1906, p. 185–270, 17 pls.; Insects and Fungi, Science Progress 1907, No. 6, 10 pp.; White Ants and Fungi. Ann. Roy. Bot. Gard. Peradenyia 1913, p. 389–393; **Sjöstdt, Y.**, *Termes Lilljeborgi*, eine neue wahrscheinlich pilzanbauende Tagtermite aus Kameroun. Festchr. Lilljeborg. Upsala 1896, p. 267–280, 1 pl.; Monographie der Termiten Afrikas *l.c.* 1900–1904; **Smeathman, H.**, Some Account of the Termites, etc. *l.c.* 1781; **Smith, C. F.**, White Ants as Cultivators of Fungi. Amer. Nat. 30, 1896, p. 319; **Sykes, M. L.**, Termites and Ants of West Africa. Trans. Manchester Micr. Soc. 1899, p. 85–91; **Uichanco, L.**, General

Facts in the Biology of Philippine Mound-building Termites. *l.c.* 1919, **Wheeler, W. M.**, The Fungus-Growing Ants of North America. Bull. Amer. Mus. Nat. Hist. 23, 1907, p. 669–807 (fungus-growing termites, p. 775–786).

9. The more important recent papers on termitophiles may be found in the following bibliography: **Assmuth, J.**, *Termitoxenia assmuthi* Wasm. Anatomisch-histologische Untersuchung. Inaug. Dissert. Berlin 1910, 53 pp.; **Brues, C. T.**, Some Stages in the Embryology of Certain Degenerate Phoridæ and the supposed hermaphroditic genus Termitoxenia. Science N. S. 27, 1908, p. 942; **Breddin** and **Börner,** Ueber *Thaumatoxena Wasmanni*, den Vertreter einer neuen Unterordnung der Rhynchoten. SB. Ges. Naturf. Freunde Berlin, 1904, p. 84–93; **Bugnion, E.**, Termitoxenia, Étude Anatomo-histologique. Ann. Soc. Ent. Belg. 57, 1913, p. 23–44, 3 pls.; **Hozawa, S.**, Note on a New Termitophilous Coleoptera found in Formosa (*Ziælus formosanus*). Annot. Zool. Japon. 8, 1914, p. 483–488, 1 pl.; **Schiödte, J. C.**, Corotoca og Spirachtha, Staphyliner some föde levende Unger, og ere Huusdyr hos en Termit. Copenhagen 1854, 19 pp., 2 pls.; Observations sur les Staphylines vivipares, qui habitent chez les termites, etc. Ann. Sc. Nat. Zool. (4) 5, 1856, p. 169–183, 2 pls.; **Schwarz, E. A.**, Staphylinidæ Inquilinous in the Galleries of *Termes flavipes.* Amer. Ent. 3, 1880, p. 15; Termitophilous Coleoptera found in North America. Proc. Ent. Soc. Wash. 1, 1889, p. 160–161; **Silvestri, F.**, Contribuzione alla Conoscenza dei Termitidi e Termitofili dell Eritrea. Redia 3, 1905, p. 341–359, 22 figs.; Descrizione di nuovi Polydesmoideæ termitofile. Zool. Anzeig. 38, 1911, p. 486–492, 4 figs.; Termitofili raccolti dal Prof. K. Escherich a Ceylon Zool. Jahrb. Syst. 30, 1911, p. 401–418, 7 pls.; Contribuzione alla Conoscenza dei Termitidi e Termitofili del' Africa Occidentale *l.c.* 1914–20; Descrizione di alcuni Staphylinidæ termitofile delle regioni orientale e australiane. Bol. Lab. Zool. Scuol. Agric. Portici 15, 1921, p. 3–23, 14 figs.; **Trägardh, I.**, Description of Termitomimus, a new genus of termitophilous physogastric Aleocharini, with notes on its anatomy. Zool. Stud. tilläg. Prof. T. Tullberg. Upsala 1907, 1 pl.; Notes on a termitophilous Tineid Larva. Ark. Zool. 3, 1907, 7 pp.; Contributions to the Knowledge of Thaumatoxena Bredd. and Börn. *ibid.* 4, 1908, 12 pp., 7 figs.; Cryptopteromyia, eine neue Phoriden-Gattung mit reduzierten Flügeln aus Natal; nebst Bemerkungen über Thaumatoxena Br. et Börn. Zool. Jahrb. Syst. 28, 1909, p. 329–348, 1 pl., 16 figs.; **Warren, E.**, Termites and Termitophiles. S. Afr. Journ. Sc. Cape Town. Sept. 1919; Observations on the Comparative Anatomy of the Termitophilous Aliocharine *Paracorotoca Akermanni* (Warren). Ann. Natal Mus.

4, 1920, p. 297–366, 6 pls.; **Wasmann, E.,** Vergleichende Studien über Ameisen-und Termitengäste. Tijdschr. Ent. 23, 1890, p. 27–97, 1 pl.; Neue Termitophilen, mit einer Uebersicht über die Termitengäste. Verh. zool.-bot. Ges. Wien. 1891, p. 647–658, 1 pl.; Kritisches Verzeichnis der myrmekophilen und termitophilen Arthropoden. Berlin, F. L. Dames, 1894; Die Ameisen und Termitengäste von Brasilien. Verh. zool.-bot. Ges. Wien. 1895, p. 137–179; Die Myrmecophilen und Termitophilen. C. R. 3ᵐᵉ Congr. Intern. Zool. Leyden 1896, p. 410–440; Neue Termitophilen und Termiten aus Indien. Ann. Mus. Civ. Genova (2) 16, 1896, p. 613–630; Einige neue termitophile Myrmedonien aus Birma. *Ibid.* (2) 18, 1897, p. 28–31; Neue Termitophilen und Myrmecophilen aus Indien. Deutsch. Ent. Zeitschr. 1899, p. 145–156, 2 pls.; Ein neuer Termitodiscus aus Natal. *ibid.* 1899, p. 401; G. D. Haviland's Beobachtungen über die Termitophilie von *Rhopalomelas angusticollis* Boh. Verh. zool.-bot. Ges. Wien. 1899, p. 245–249; Zur Kenntnis der termitophilen und myrmecophilen Cetoniiden Südafrikas. Ill. Zeitschr. Ent. 5, 1900, p. 65–67, 81–84; Termitoxenia, ein neues flügelloses physogastres Dipterengenus aus Termitennestern. 1. Teil. Zeitschr. wiss. Zool. 67, 1900, p. 599–617, 1 pl.; 2. Teil, *ibid.* 70, 1901, p. 289–298; Zur näheren Kenntnis der termitophilen Dipterengattung Termitoxenia. Verh. 5. Intern. Zool. Kongr. Jena. 1902, p. 852–872, 1 pl.; Species novæ insectorum termitophilorum ex America meridionali. Tijdscher. Ent. 65, 1902, p. 95–107, 1 pl.; Species novæ insectorum termitophilorum a Dr. F. Silvestri in America meridionali inventæ. Boll. Mus. Zool. R. Univ. Torino 17, 1902, 6 pp.; Termiten, Termitophilen und Myrmekophilen gesammelt auf Ceylon von Dr. Horn 1899. Zool. Jahrb. Syst. 17, 1902, p. 99–164, 2 pls.; Die Thorakalanhänge der Termitoxeniidæ usw. Verh. deutsch. Zool. Ges. 1903, p. 113–120, 2 pls.; Zur näheren Kenntnis des echten Gastverhältnisses (Symphilie) bei den Ameisen und Termitengästen. Biol. Zentralbl. 23, 1903, p. 261–276, 298–310; Neues über die zusammengesetzten Nester und gemischten Kolonien der Ameisen. Allg. Zeitschr. Ent. 6, 1902, 7, 1903; Termitophilen aus dem Sudan (Collab. Forel, Breddin, Escherich). Res. Swed. Zool. Exped. Egypt. Part 1, 1903, 1 pl.; Die phylogenetische Umbildung ostindischer Ameisengäste und Termitengäste. C. R. 6. Congr. Intern. Zool. Berne 1904, 1905, p. 436–449, 1 pl.; Beispiele rezenter Artenbildung bei Ameisen und Termitengästen. Festschr. Rosenthal, Leipzig 1906, p. 44–58; Biol. Zentralbl. 26, p. 565–580; Termitusa, nouveau genre d'Aléochariens termitophiles. Rev. d'Ent. 25, 1905, p. 199–200; Die moderne Biologie und die Entwicklungstheorie. 2 Aufl. Freiburg i/Br. 1906, pp. XXX + 530, 7 pl., 54 textfigs.; Termitophilen in Schultze, Forschungsreise Westzentr. Südafrik.

1908, p. 239–243; Deutschr. Mediz. naturw. Ges. Jena 13, 1908, p. 441–445, 1 pl.; Termitophile Coleopteren aus Ceylon (in Escherich's Termitenleben auf Ceylon, p. 231–232); Zur Kenntnis der Termiten und Termitengäste vom belgischen Congo. Rev. Zool. Afr. 1, 1911, p. 91–117, 145–176, 8 pls.; Tabelle der Termitophya- und Xenogaster-Arten, Zool. Anzeig. 38, 1911, p. 428–429; Neue Beiträge zur Kenntnis der Termitophilen und Myrmecophilen. Zeitschr. wiss. Zool. 101, 1912, p. 70–115, 3 pls.; Revision der Termitoxeniinæ von Ostindien und Ceylon, Ann. Soc. Ent. Belg. 57, 1913, p. 16–22, 2 figs.; Neue Beispiele der Umbildung von Dorylinengästen zu Termitengästen. Verh. Ges. Deutsch. Naturf u. Aertzte 2, 1912, p. 254–257; Wissenschaftliche Ergebnisse einer Forschungsreise nach Ostindien. 5. Termitophile und myrmecophile Coleopteren gesammelt von Prof. Dr. v. Buttel-Reepen 1911–1912. Zool. Jahrb. Syst. 39, p. 169–210, 2 pls.; Das Gesellschaftsleben der Ameisen I Bd.; Münster i. W. 1915; Myrmecophile und termitophile Coleopteren aus Ostindien gesammelt von P. J. Assmuth und B. Corporael. 1 Teil, Tidschr. Ent. 60, 1917, p. 382–408; 2 Teil, Wien. Ent. Zeitg. 37, 1918, p. 1–23, 3 pls.; **Wheeler, W. M.**, A Study of Some Ant Larvæ, with a Consideration of the Origin and Meaning of the Social Habit among Insects. Proc. Amer. Phil. Soc. 57, 1918, p. 293–343, 12 figs.

10. **Ottramare, J. H.**, Quelques reflexions à propos de l'action de l'obsurité sur les êtres vivantes. C. R. Soc. Biol. Paris 82, 1919, p. 190–191.

11. **Bohn** and **Drzwina**, La Biologie Générale et la Psychologie Comparée. Rev. Scient. 1912.

12. **Loeb, J.**, The Blindness of the Cave Fauna and the Artificial Production of Blind Fish Embryos by Heterogeneous Hybridization and by Low Temperatures. Biol. Bull. 29, 1915, p. 50–67, 13 figs.

13. The following literature deals with the infusorial intestinal parasites of termites: **Brunelli, G.**, Sulla destruzione degli Oociti nelle regine dei Termitidi infetti da Protozoi. Rend. Accad. Lincei 2, 1905, p. 718–721, 1 fig.; **Bugnion, E.**, and **Popoff, N.**, Les Calotermes de Ceylan. Mém. Soc. Zool. France 23, 1910, p. 124–144, 3 pls., 2 figs.; Le Termite a Latex de Ceylan, *Coptotermes travians* Haviland. *ibid.* 23, 1910, p. 105–122, 2 pls., 2 figs.; **Buscaloni, L.**, and **Comes, S.**, La digestione delle membrane vegetali per opera dei Flagellati contenuti nell' intestino dei Termitidi e il problema del simbiosi. Atti Accad. Gioenia Sc. Nat. Catania (5) 3, 1910, p. 1–16, 4 figs.; **Comes, S.**, *Lophophora vacuolata.* Nuova Genere e Nuova Specie di Flagellato dell' Intestino dei Termitidi. Bull. Acad. Gioenia Sc. Nat. Catania (2) 13, 1910, p. 11–19, 3 figs.; Alcune Considerazione al proposito del

Dimorfismo Sessuale riscontrato in *Dinenympha gracilis* Leidy. *ibid.*
(2) 13, 1910, p. 20–29, 7 figs.; Notizie sulla Morfologia e Reproduzione
di *Monocercomonas termitis. ibid.* (2) 3, 1914, p. 15–27, 5 figs.;
De Mello, F., The Trichonymphal Parasites of Some Indian Termites.
Rep. Proc. 3rd Ent. Meet. Pusa, Feb. 1919; Calcutta 3, 1920, p. 1009–
1022, 3 pls.; **Dobell, C. C.**, On Some Parasitic Protozoa from Ceylon.
Spolia Zeylandica 7, 1910, p. 65–87, 1 pl.; **Doflein, F.**, Die Protozoen als
Parasiten und Krankheitserreger. Jena. 1901; **Foa, A.**, Recerchi
sulla Reproduzione dei Flagellati. 2. Processo di Divisione delle Tri-
coninfa. Atti Accad. Lincei (5) 13, p. 618–625, 5 figs.; **Frenzel, J.**,
Leidyonella cordubensis nov. gen. nov. sp. Eine neue Trichonymphide.
Arch. mikr. Anat. 38, 1891, p. 301–316, 4 figs.; **Grassi, B.**, Intorno
ad alcuni protozoi parassiti delle Termiti. Atti Accad. Gioenia (3) 18,
1885, p. 235–240; Flagellati viventi nei Termiti. Mem. R. Accad.
Lincei (5) 12, 1917, p. 1–68, 10 pls.; **Grassi, B.**, and **Foa, A.**, Ricerche
sulla Riproduzione dei Flagellati I. Processo di Divisione delle Joenia
e Forme Affini. Rend. Accad. Lincei 13, 1904, p. 241–255, 17 figs.;
Intorno ai Protozoi dei Termitidi. Nota Prelim. *ibid.* (5) 20, 1911,
p. 725–741; **Hartmann, M.**, Untersuchungen über Bau und Entwicklung
der Trichonymphiden (*Trichonympha hertwigi* n.sp.). Festschr. R.
Hertwig. 1, 1910, p. 339–396, 4 pls., 3 figs.; **Imms, A. D.**, On the Structure and Habits of Archotermopsis, etc., *l.c.* 1919; **Janicki, C.**, Zur
Kenntnis des Parabasalapparatus bei parasitischen Flagellaten. Biol.
Centralbl. 31, 1911, p. 321–330, 8 figs.; Untersuchungen an parasitischen
Flagellaten II. Die Gattungen Devescovina, Parajoenia, Stephano-
nympha, Calonympha. Zeitschr. wiss. Zool. 112, 1915, p. 573–689,
6 pls., 17 figs.; **Kent, W. S.**, Infusorial Parasites in White Ants in
Tasmania. Pap. Proc. R. Soc. Tasmania for 1884, 1885, p. 270–273;
Ann. Mag. Nat. Hist. (5) 15, 1885, p. 450–453; **Kofoid, C. A.**, and
Swezy, O. S., Mitosis and Mutiple Fission in Trichomonad Flagellates.
Proc. Amer. Acad. Arts Sc. 51, 1915, p. 289–378, 8 pls., 7 figs.;
Studies of the Parasites of Termites, I to IV. Univ. Cala. Publ.
Zool. 20, 1919, pp. 1–116, 14 pls., 8 figs.; Flagellate Affinities of Tri-
conympha. Proc. Nation. Acad. Sc. 5, 1919, p. 9–16; **Leidy, J.**, On
Intestinal Parasites of *Termes flavipes* Proc. Acad. Nat. Sc. Phila.
1877, p. 146–149; The Parasites of Termites. Journ. Acad. Nat. Sc.
Phila. (2) 8, 1874–81, p. 425–447; **Porter, J. F.**, Trichonympha and
Other Parasites of *Termes flavipes*. Bull. Mus. Comp. Zool. 31, 1897,
p. 47–68, 6 pls.

14. **Demoor, J., Massart, J.**, and **Vandervelde, E.**, Evolution by Atro-
phy in Biology and Sociology. Transl. by Mrs. C. Mitchell. Intern.
Sc. Ser. New York, Appleton 1899.

INDEX OF SUBJECTS

357

INDEX OF AUTHORS

INDEX 375